Basic Engineering Thermodynamics

P. B. Whalley

Department of Engineering Science
University of Oxford

Oxford   New York   Melbourne
OXFORD UNIVERSITY PRESS

Oxford University Press, Great Clarendon Street, Oxford, OX2 6DP
Oxford New York
Athens Auckland Bangkok Bogota Bombay Buenos Aires
Calcutta Cape Town Dar es Salaam Delhi Florence Hong Kong
Istanbul Karachi Kuala Lumpur Madras Madrid Melbourne
Mexico City Nairobi Paris Singapore Taipei Tokyo Toronto
and associated companies in
Berlin Ibadan

Oxford is a trade mark of Oxford University Press

Published in the United States by
Oxford University Press Inc., New York

First published 1992
Reprinted 1995 (with corrections), 1997

A catalogue record for this book is available from the British Library

Library of Congress Cataloging in Publication Data
(Data applied for)
ISBN 0 19 856255 1 (Pbk)

Printed and bound in Great Britain by
Biddles Ltd, Guildford and King's Lynn

Preface

The objective of this book is to give a brief introduction to thermodynamics and its application to engineering. Most textbooks are too long and contain too much material. Almost every topic in this book could be expanded greatly: I have tried to resist this temptation. Many thermodynamics textbooks adopt a rigorous approach to the subject. I have tried to produce a more informal introduction— giving illustrative examples where possible. Most of the material in this book started life as a course of lectures in thermodynamics to first year engineering students at Oxford, although some of the more applied material later in the book is actually taught in the second year.

As well as decisions about what material to cover and what approach to take, an author has more technical decisions to make. The most difficult one, for me, was the sign convention for work transfer in thermodynamic processes. Traditionally in the UK, engineers have used the convention that the work done **by** a system is positive. However, chemists and physicists have considered that positive work is work transfer **to** the system. I have followed the traditional, but now increasingly challenged, view that positive work is work done **by** the thermodynamic system.

I have assumed that readers have some knowledge of partial differentiation and its notation, but otherwise little advanced mathematics is required.

I wish to acknowledge that extracts from steam tables given in Chapters 4, 6, 7, and 10 are reproduced by permission from 'Engineering Tables and Data' by my colleagues A.M. Howatson, P.G. Lund, and J.D. Todd (second edition, 1991), published by Chapman and Hall. I also wish to thank my colleagues at Oxford for their helpful criticisms of my lecture notes and of this text, and for permission to adapt some of the problems set to first year undergraduates for use in this book. I also have a great debt to those undergraduates who have pointed out many of my mistakes. Finally, the staff of Oxford University Press have been unfailingly helpful: it has been a pleasure to work with them.

Those who wish to study thermodynamics in depth will soon exhaust this book. For them the list of further reading below may be useful.

- For a much more thorough treatment of engineering thermodynamics, see 'Engineering Thermodynamics—Work and Heat Transfer' by G. Rogers and Y. Mayhew, published by Longman Scientific and Technical. The fourth edition (1992) does not use the sign convention for work which I have here, and which was used in the well-used third edition of Rogers and Mayhew.

- For a concise treatment of the application of thermodynamics to chemical systems, see 'Basic Chemical Thermodynamics' by E.B. Smith, published by Oxford University Press, fourth edition, 1990. It was seeing Brian Smith's elegant book that first inspired me to write this book.

- For an account of thermodynamics applied to problems in physics, see 'Heat and Thermodynamics' by M.W. Zemansky and R.H. Dittman, published by McGraw Hill, 1981. It was an earlier version of this book which first kindled my interest in thermodynamics as an undergraduate.

- For an account of thermodynamics and its connections with kinetic theory and statistical mechanics see, 'Thermodynamics, Kinetic Theory, and Statistical Thermodynamics' by F.W. Sears and G.L. Salinger, published by Addison-Wesley, third edition, 1975.

- Finally if thermodynamics ever seems a dull subject, turn to 'Engines, Energy, and Entropy' by J.B. Fenn, published by W.H. Freeman & Co., 1982. This is probably the only thermodynamics book which needs no knowledge of mathematics, and also includes the cartoon adventures of Charlie the Caveman.

Department of Engineering Science,   P.B.W.
University of Oxford
June 1992

Many thanks to the following who pointed out misprints in the first printing: Sue Johnson, Victor Loo, Francis Plunkett, and Hans Wesselingh.

January 1995   P.B.W.

Contents

1
Introduction

1.1 Key points of this chapter

- Engineering thermodynamics is mainly concerned with heat and work, and their interconversion. Heat and work are both measured in Joules. (Section 1.2)

- Mechanical work is the scalar product of two vectors: force and distance. (Section 1.3)

- Heat is much harder to define, but when heat is added to a body the temperature rise is usually proportional to the heat added. (Section 1.4)

- The zeroth law of thermodynamics says that if two bodies are each in thermal equilibrium with a third body, then the two bodies themselves will be in thermal equilibrium. This makes it possible to measure temperature consistently. (Section 1.5)

- Heat and work are now recognised as different, but not equivalent, forms of energy following the work of Rumford and Joule. (Section 1.6)

- If the mass, the specific heat, and the temperature rise of a body are known, then the heat transferred to the body can be found. (Section 1.7)

- The main applications of engineering thermodynamics are briefly reviewed, and references to the rest of the book are given. (Section 1.8)

1.2 What is thermodynamics?

Thermodynamics is a subject which is used in many branches of science and is concerned with heat and work, and the conversion of heat into work and vice versa. We shall see that although work is easily and completely converted into heat, the converse is not true.

Thermodynamics is also concerned with the reversibility of processes: a change might be accomplished but can that change by entirely reversed? Thermodynamics is also concerned with the efficiency of processes. We shall see that there are good reasons why electricity generating stations convert what appears to be a rather low proportion of the heat energy liberated by burning coal or oil into electrical energy. Thermodynamics is also concerned with chemical reactions and their equilibrium, though the application of thermodynamics to chemical reactions is outside the scope of this book, as is the application to geology and to the attainment of very low temperatures.

In this book we are concerned with the fundamental quantities of heat and work. We are now so used to measuring these in the same units—the SI unit is the Joule—that we forget that it took many years of work and careful experimentation to convince sceptics that heat and work really were different manifestations of one thing—energy.

Before describing some of these experiments it is useful to have a reminder of what we mean by the terms 'work' and 'heat'.

1.3 The nature of work

Mechanical work is simply defined as the product of force and distance moved by the point of application of the force. Strictly we must remember that both force and distance are vector quantities, and the particular product we want is the scalar or dot product. Thus writing W for the work (J), $\mathbf{F}$ for the force (N), and $\mathbf{x}$ for the distance (m), then:

$$W = \mathbf{F} \cdot \mathbf{x} \tag{1.1}$$

If $\mathbf{F}$ and $\mathbf{x}$ are at right angles the work done is zero.

1.4 The nature of heat

Heat is less easily defined, but we do know that if we add heat to a mass of substance, then in general the substance heats up, that is the temperature increases. At a molecular level the atoms or the molecules in the substance are moving more quickly. Denoting the amount of heat by Q (J), and the temperature rise ΔT (measured in some appropriate units), then the variables are linked by:

$$Q \propto \Delta T \qquad (1.2)$$

Later, in Section 1.7, the necessary extra variables will be introduced to turn this into a true equation enabling Q to be calculated from ΔT.

However, a definition of heat in terms of eqn 1.2 is not always satisfactory. For example, when heat is added to water at its boiling temperature, the effect of the addition of heat is not to cause the temperature to increase but to cause some of the water to be turned into steam. An alternative definition of heat is that it is what flows from a hot object to a cold object when they are brought into contact. This definition leads to the concept of temperature. The two objects have the same temperature if no heat flows between them when they are brought into contact.

1.5   The zeroth law of thermodynamics

Closely allied to this concept of temperature is the **zeroth law of thermodynamics** first formulated in 1931[1]. This is that if two objects (A and B) are both in thermal equilibrium with a third object C, then A and B will be in thermal equilibrium. Thermal equilibrium here means that if the objects are brought into contact then no heat will flow between them: this implies that the objects have the same temperature. Thus, using the symbol T for temperature, the zeroth law implies that if:

$$T_A = T_C \qquad (1.3)$$

and

$$T_B = T_C \qquad (1.4)$$

then

---

[1] The zeroth law is so named because it was formulated well after both the first and second laws but is a necessary precursor to these laws. The date of first formulation of the laws of thermodynamics is worthy of note:

Zeroth	1931
First	1850
Second	1824

Table 1.1   Simple definition of practical temperature scales

	Freezing point	Boiling point
Celsius	0°	100°
Fahrenheit	32°	212°
Réaumur	0°	80°

$$T_A = T_B \tag{1.5}$$

Written algebraically in this form, the zeroth law appears obviously true: eqns 1.3 to 1.5 form a consistent set, in other words eqn 1.5 certainly follows from eqns 1.3 and 1.4. However, we are assuming that temperature is a well-behaved, conventional variable, and that the equals sign has its normal meaning in the equations. Suppose in eqns 1.3 to 1.5 the A, B, and C were three different chemical substances and the equals sign meant that a chemical reaction took place. It is quite clear that if both A and B react with C, then it is not necessarily true that A and B will react with each other. The zeroth law, like the other laws of thermodynamics we shall meet, is not provable. We have to take it on trust as true, and we shall find no impossible consequences as a result of our assumption.

One consequence of the zeroth law is that temperature can be measured consistently. If thermometers A and B are both used to measure the temperature of water C, the thermometers will register the same temperature. The scales of temperature in common use are the Celsius scale (formerly called Centigrade), the Fahrenheit scale, and (less commonly now) the Réaumur scale. On these scales the freezing and boiling points of water at atmospheric pressure are shown in Table 1.1. Later we will meet more precise definitions of temperature scales—both theoretical and practical.

1.6   Heat and work as different forms of energy

As stated in Section 1.2 above, we now recognize that heat and work are just different forms of energy by using the same unit, the Joule in SI units, for both heat and work. However, this took a long time to be established, and it was not until about 1850 before this was fully accepted.

Before the work of Joule many scientists believed the **caloric theory of heat**. Heat or caloric fluid was held to be a weightless fluid associated with matter. If the matter is heated, extra caloric fluid is added to the matter and the matter expands. Other thermal phenomena could also be explained by the theory. For example, the flow of heat through a solid was thought of as the flow of caloric fluid through the solid. Different solids allowed the caloric fluid to move more or less easily, thus explaining the very different thermal conductivities of solids. Although believed by many eminent scientists even up to the middle of the nineteenth century, others had doubts. One noted opponent of the caloric theory was Benjamin Thompson (who became Count Rumford). He performed a famous experiment which consisted of attempting to bore out the central hole in a cannon barrel by using a blunt boring tool. The cutting tool was so blunt that very little metal was removed but a lot of heat was generated. Rumford maintained that the large amount of heat could not have come from the small amount of metal removed. However, proponents of the theory turned the experiment to their advantage by maintaining that it just demonstrated that there was a very large amount of caloric fluid in the metal. The inefficient boring process had just resulted in a large amount of caloric fluid being squeezed out of the metal. Hence the large amount of heat generated was quite explicable.

Experiments of the type where heat was added, and the matter weighed before and after the process, and no change was found led to the belief that the caloric fluid was weightless. Another experiment which challenged the caloric theory was performed in 1799 by Davy. He showed that if two blocks of ice are rubbed together, some of the ice melts.

It was the experiments of Joule which led to the realization that the caloric theory was not correct. Joule performed a series of experiments between 1843 and 1850 where falling weights stirred a liquid (see Fig. 1.1). When the experiment was repeated many times a temperature rise in the liquid could be measured. The important point about Joule's experiment was that however the work on the liquid was done, and whatever the liquid was (he used water, mercury, and whale oil) the same result was obtained. One unit of heat was always equal to the same number of units of work. In the units used by Joule, 1 British Thermal Unit (the amount of heat necessary to raise the temperature of one pound of water by one degree Fahrenheit) was equal to 772.5 foot pounds of work. Joule's results were, for their time, remarkably accurate. The most accurate modern measurements produce an only slightly revised figure of 778 foot pounds. Joule had shown that heat and work were, in some sense, equivalent. Work could certainly be converted into an equivalent amount of heat. Later it will be seen that the conversion of heat into work is not so

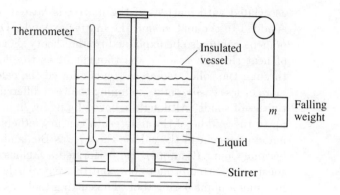

Fig. 1.1 Joule's experiment to measure the amount of heat generated from a given amount of mechanical work

simple. Heat cannot be converted into an equivalent amount of work: only some of the heat can be converted into work.

1.7 More about heat

Equation 1.2 is not a true equation. It tells us only that the heat is proportional to the temperature rise. It is no surprise that for a given temperature rise ΔT the heat Q is proportional to the mass m of the material to which the heat is added. Therefore:

$$Q = mC\Delta T \tag{1.6}$$

The constant of proportionality C in eqn 1.6 has units of J/kg °C and is the specific heat of the substance. The specific heat is a property of the substance but can (and does) vary with the temperature of the substance. Later we shall see that gases have two principal specific heats, but liquids and solids can be regarded for practical purposes as having a single specific heat.

Equation 1.6 can be used to calculate the heat added to a mass in many, but not all, circumstances. As pointed out in Section 1.4 it will not work when a change of phase is taking place.

1.8 Engineering applications of thermodynamics

In this book the early chapters are inevitably concerned with an introduction to the essential elements of thermodynamics. It is sometimes easy to forget the engineering applications of this theory. The main applications are summarized very briefly here. References to later sections in the book are also given.

1. Thermodynamics allows the unambiguous definition of temperature: first as a by-product of the behaviour of ideal gases (see Section 2.5), then as a purely empirical scale based on the freezing or boiling points of pure substance (again see Section 2.5), and finally, without reference to the properties of any substance as a consequence of the second law of thermodynamics (see Section 9.3).

2. Thermodynamics characterizes changes in the pressure, volume, or temperature of a substance, a thermodynamic process, as possibly being carried out in two distinct ways: reversibly or irreversibly (see Section 5.3). Reversible processes are often seen as an unattainable ideal. The theory for a reversible process will often, for example, set a maximum value to the amount of mechanical work that can be extracted from a particular change.

3. The heart of engineering thermodynamics is the interconversion of heat and work. Although these are in one sense just different forms of energy measured in the same units (see Section 1.6), they are significantly different because whilst work may be completely converted into heat, heat cannot be completely converted into work. This is, in fact, an informal statement of the second law of thermodynamics (see Section 8.4). A generalized view of a machine to convert heat partially into work is presented: the heat engine (see Section 8.2). This takes in heat at a high temperature, converts some of the energy into useful mechanical work, and then rejects heat at a low temperature. It is proved that the heat engine which converts the greatest fraction of the heat input into work, the most efficient heat engine, is entirely made up of reversible processes (see Section 9.2), and that the efficiency of this engine depends only on the temperature at which the heat is taken in and the temperature at which it is rejected (see Section 9.3). A heat engine can be run in the reverse direction to take in heat at a low temperature and reject it at a high temperature. To do this work has to be added to the cycle, not extracted from it. This machine is a refrigerator or a heat pump (see Section 8.3).

4. Heat engines are discussed in more detail in a number of places.

   (a) A closed system using an ideal gas is defined in Section 6.8.

(b) The steady flow Rankine cycle operating with steam and water is used for large-scale power production in electricity generating stations (see Section 7.7 and Chapter 14).

(c) Chapter 15 describes the gas turbine cycle for the conversion of heat into work. This cycle is used in aircraft engines and in electricity generating stations designed to produce power a short time after start up.

(d) Section 15.4 describes a combination of the Rankine cycle and the gas turbine cycle: the combined cycle, which is now coming into use has a very high efficiency.

(e) Chapter 17 briefly describes internal combustion cycles which again convert heat (in the fuel which is burnt in the engine) into useful mechanical work. The main types of cycle are the petrol or Otto cycle (see Sections 17.2 and 17.3), and the Diesel cycle (see Section 17.4).

5. Practical heat pump and refrigerator cycles are discussed in Chapter 16.

6. Many of the practical cycles use turbines and compressors. A turbine extracts work from a high pressure gas or vapour. A compressor raises the pressure of a gas or vapour. Chapter 13 describes the main thermodynamic aspects of the operation of these machines, but the fluid mechanics of the flows in the machines is not discussed in this book. Turbines are also discussed briefly in Section 7.2.

7. Much of the manipulation of thermodynamic equations in this book is delayed until Chapter 18. Maxwell's equations are very useful for the proof of thermodynamic equations, and Bernouilli's equation is proved in a very simple way (see Section 18.5). Bernouilli's equation is one of the fundamental equations of fluid mechanics. Under a fairly tightly constrained range of circumstances it relates the pressure in a fluid (either a liquid or a gas) to the fluid velocity.

1.9 Problems

1.1 Find the temperature at which the numerical values of the temperature in degrees Centigrade and Fahrenheit are equal.

1.2 In Fig. 1.1 the falling weight has mass 10 kg and falls through a distance of 3 m. Half the potential energy is converted into heat. The liquid in the vessel is water, and the mass of the water is 5 kg. Calculate how many times the weight must be allowed to fall to raise the temperature of the water by 15 K. The specific heat of water is 4200 J/kg K.

1.3 A waterfall is 807 m high. If all the potential energy is converted into heat, calculate the rise in the water temperature.

1.4 If the rise in the water temperature in question 1.3 is actually measured, the experimental value is almost certainly much less than the calculated value. Suggest the reason for this.

1.5 An electric kettle contains 1 kg of water at 20°C. If the thermal capacity of the kettle itself is equivalent to another 0.5 kg of water, calculate the time required to bring the water to the boiling temperature if the power of the heating element is 2.5 kW.

1.6 The kettle in question 1.5 is accidentally switched on with no water covering the heating element. If the thermal capacity of the element is equivalent to 0.1 kg of water and it is assumed that the element looses no heat from its surface, calculate the time required for the element to reach 500°C.

2

Ideal gases

2.1 Key points of this chapter

- The main relevant points concerning atoms, simple molecules, molecular weight (or relative molar mass) and the kilogram-mole are reviewed. (Section 2.2)

- Ideal gases obey Boyle's law and Charles' law. For real gases these equations become more accurate at low pressure and high temperature. (Section 2.3)

- The ideal gas equation can be expressed in terms of mass or in terms of the number of kilogram-moles. (Section 2.4)

- The universal gas constant per kilogram-mole and the gas constant per unit mass. (Section 2.4)

- The ideal gas equation can be used as the basis of a method of determining temperature using the constant volume gas thermometer. (Section 2.5)

- Temperature is actually measured with reference to a number of defined points like the freezing point of liquid gold. This is the international practical temperature scale. (Section 2.5)

2.2 Atoms, molecules, and molecular weight

It is the intention here to provide only an outline of some of the important results about atoms and molecules. The primary result is that all matter is made up of molecules which in turn are made up of groups of atoms. Matter can be subdivided repeatedly until one molecule is reached. Thus common salt (sodium chloride) crystals can in principle be broken into smaller and smaller fragments until one molecule of sodium chloride is obtained. This molecule is made up of one atom of sodium and one atom of chlorine. Sodium and chlorine are elements: they consist of atoms. Sodium chloride is a compound: it consists of molecules which in turn are made up of atoms.

The picture of the atom taken here is one from around 1940 when it was known that there were three important elementary particles which went to make up every atom: protons, neutrons, and electrons.

1. The protons and neutrons form the nucleus of the atom, and have approximately the same mass (approximately 1.67×10^{-27} kg): the electron is much lighter: its mass is 9.11×10^{-31} kg.

2. The proton is electrically charged positive and the electron has a negative charge of the same magnitude. The protons and the neutrons are concentrated together in the nucleus and the electrons orbit around the nucleus. To make the atom electrically neutral the number of protons present must equal the number of electrons.

3. The number of neutrons present is always at least as large as the number of protons, except for the special case where there is one proton—in this case there can be no neutrons.

4. The number of protons determines the chemical element. Atoms of the same element can, sometimes, have different numbers of neutrons—these are different **isotopes** of the same element.

5. If the proton and the neutron are assigned a mass of one 'atomic mass unit'[1] then, because the mass of the electron is very small, the mass of each atom, in atomic mass units, is the sum of the number of protons and the number of neutrons. This mass is commonly called the mass number A. The atomic number Z is the number of protons in the atom. The mass number and atomic number of some atoms is shown in Table 2.1. Deuterium is a naturally occurring isotope of hydrogen: the hydrogen nucleus consists of a single proton, the deuterium nucleus has a neutron as well as a proton. Natural hydrogen contains approximately one part deuterium in six thousand.

6. Some important elements do not occur naturally as isolated atoms: hydrogen, nitrogen, and oxygen all exist at normal conditions

[1]There is a much more precise definition of the atomic mass unit in terms of the mass of the isotope of carbon which has six protons and six neutrons.

Table 2.1   Atomic numbers and mass numbers of some important elements

Atom	Chemical symbol	Atomic number Z	Mass number A
hydrogen	H	1	1
deuterium	D	1	2
helium	He	2	4
carbon	C	6	12
nitrogen	N	7	14
oxygen	O	8	16
argon	Ar	18	40

Table 2.2   Relative molar masses of some common elements and compounds

Name	Chemical formula	Accurate relative molar mass M	Approximate relative molar mass M
hydrogen	H_2	1.008	1
deuterium	D_2	2.014	2
helium	He	4.003	4
carbon	C	12.011	12
water	H_2O	18.017	18
'heavy' water	D_2O	20.028	20
nitrogen	N_2	28.014	28
carbon monoxide	NO	28.010	28
oxygen	O_2	31.998	32
argon	Ar	39.948	40
carbon dioxide	CO_2	44.009	44

as molecules consisting of two atoms linked together. Thus the chemical formulae for these elements are H_2, N_2, and O_2.

7. The relative molar mass (or the molecular weight for a compound or the atomic weight for an element which exists as single atoms like carbon) is the average mass of the molecule or the atom in atomic mass units. An average is necessary to take account of the relative abundance of various isotopes. The relative molar masses, M, of some common elements and compounds are shown in Table 2.2.

8. The kilogram-mole of a substance is a mass of that substance where the number of kilograms is equal to the relative molar mass (molecular weight). Thus one kilogram-mole of water is 18 kg of water, and one kilogram-mole of carbon dioxide is 44 kg of carbon dioxide. A kilogram-mole is abbreviated to kg-mole, kg-mol, kmol, or sometimes simply to mole.

9. It is a consequence of the above information that one kg-mole of any substance contains the same number of molecules. This

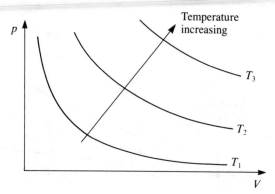

Fig. 2.1 Illustration of Boyle's law: variation of pressure with volume for a fixed mass of gas at a number of temperatures

number is Avogadro's number and is 6.023×10^{26} molecules/kg-mole.

2.3 Boyle's law and Charles' law

Boyle's law states that the pressure, p, of a fixed mass of gas is inversely proportional to the volume, V, occupied by the gas at constant temperature. Expressed as an equation:

$$pV = C \tag{2.1}$$

where C depends on the temperature and the mass of gas. Careful measurements show that eqn 2.1 is only approximately true, but that it becomes more and more true at low pressure and high temperature. Equation 2.1 is plotted in Fig. 2.1 for a fixed mass of gas at different temperatures.

Charles' law states that the pressure, p, of a fixed mass of gas is linearly dependent on the temperature of the gas. Figure 2.2 shows the variation of pressure with temperature for a number of different volumes.

Figure 2.2 suggests that if the temperature were low enough the pressure would become equal to zero. This temperature is numerically equal to -273.15°C. A new temperature scale can now be used. The zero of the scale is this temperature of -273.15°C and the intervals of the scale are the same as the Celsius scale. On this scale, the absolute scale, where the temperature is measured in units named

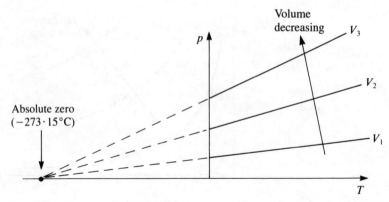

Fig. 2.2   Illustration of Charles' law: variation of pressure with temperature for a fixed mass of gas at a number of volumes

Kelvin (abbreviated K), the normal freezing point of water is 273.15 K and the normal boiling point of water is 373.15 K. Denoting the temperature of the gas measured on the absolute scale by T, then Charles' law can be expressed as an equation:

$$p/T = C \qquad (2.2)$$

where C depends on the volume and the mass of gas. Again, careful measurements show that eqn 2.2 is only approximately true, but that it becomes more and more true at low pressure and high temperature.

2.4   The ideal gas equation and the universal gas constant

It will be evident that Boyle's law and Charles' law, eqns 2.1 and 2.2, can be combined into a single equation:

$$pV/T = C \qquad (2.3)$$

where the value of C depends on the mass of the gas. It is also obvious that if the mass of the gas is doubled, then the volume V will double. Hence eqn 2.3 can be re-written as:

$$\frac{pV}{mT} = R \qquad (2.4)$$

or more usually as:

$$pV = mRT \qquad (2.5)$$

R in eqns 2.4 and 2.5 is not now dependent on p, V, T, or m but only on the composition of the gas. For example for air, R is equal

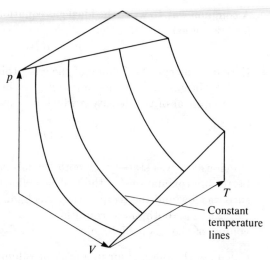

Fig. 2.3   Graphical interpretation of the ideal gas equation

to 287 J/kg K. R is known as the gas constant, and eqn 2.5 as the
ideal gas equation. In eqns 2.4 and 2.5 the units of R are as written
above; the units of p, V, m, and T are Pa, m³, kg, and K. The Pascal,
Pa (alternatively N/m²) is the SI unit of pressure: 10^5 Pa = 1 bar.

Alternatively the ideal gas equation can be written in the form:

$$pV = nR_oT \tag{2.6}$$

where n is the number of kg-moles of gas present, and R_o is the universal gas constant applicable to all gases. R_o has the value 8314 J/kg-
mole K. The two forms of the ideal gas equation (eqns 2.5 and 2.6) are
alternatives. Which is used depends on the circumstances and the information available. Both these equations are illustrated graphically
in Fig. 2.3 as a three-dimensional surface. The earlier Figs. 2.1 and
2.2 are both two-dimensional projections of this three-dimensional
surface. A consequence of eqns 2.5 and 2.6 is that:

$$R = R_o/M \tag{2.7}$$

From eqn 2.7 the gas constant for any gas can be calculated. Thus
for water vapour with $M = 18$:

$$R = 8314/18 = 461 \text{ J/kgK} \tag{2.8}$$

A common way to write the ideal gas equation is in the form of eqn 2.5 for 1 kg of gas:

$$pv = RT \tag{2.9}$$

Here v is the specific volume of the gas—this is the volume occupied by 1 kg of gas, and so the units of v are m^3/kg. Specific volume is the reciprocal of the density ρ which has units kg/m^3:

$$v = 1/\rho \tag{2.10}$$

Previously it was stated that both Boyle's law and Charles' law were better approximations to the real behaviour of gases when the pressure was low and the temperature was high. It can now be seen from eqn 2.9 that both these conditions tend to imply that the specific volume of the gas is high, and so the density of the gas is low.

Example 2.1 Calculate the mass of air in a tank of volume 0.1 m³ at a pressure of 1 bar and a temperature of 20°C.

Solution First convert the variables to SI units: $p = 10^5$ Pa and $T = 293$ K. For air the effective value of the relative molar mass, M is 29, and so using eqn 2.7: $R = 8314/29 = 287$ J/kg K. Then, as the mass has to be calculated, eqn 2.5 is used:

$$m = \frac{pV}{RT} = \frac{10^5 \times 0.1}{287 \times 293} = 0.119 \text{ kg}$$

Example 2.2 Calculate the pressure in a tank of volume 5 m³ containing 2 kg-mole of helium at a temperature of 100 K.

Solution The number of kg-moles is 2 and the gas could be any gas. As the number of moles is given it is obviously more convenient to use eqn 2.6:

$$p = \frac{nR_oT}{V} = \frac{2 \times 8314 \times 100}{5} = 3.32 \times 10^5 \text{ Pa}$$

2.5 Determination of temperature using the ideal gas equation

Equation 2.5 contains T, and so in principle it can be used to determine the temperature. There are a number of difficulties: although the pressure can be measured easily, it is more difficult to measure

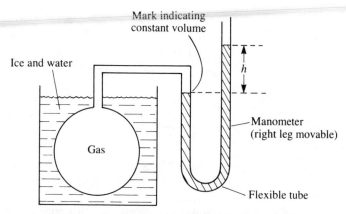

Fig. 2.4   The constant volume gas thermometer

the volume of the gas, but measuring the mass and the gas constant is not at all easy. There is then the further difficulty that eqn 2.5 is only an approximate equation. All these difficulties can be overcome by the constant volume gas thermometer (see Fig. 2.4). When taking a reading the manometer is adjusted so that the manometer fluid is always at a mark on one leg of the manometer. This keeps the volume of the gas constant. The bulb containing the gas is then exposed to two different temperatures and the pressure of the gas is measured with the manometer. If one of these temperatures is the temperature of melting ice T_i (273 K) and the corresponding pressure is p_i, and if the temperature to be measured and the corresponding pressure are T and p, then:

$$p_i V = mRT_i \tag{2.11}$$

and

$$pV = mRT \tag{2.12}$$

Eliminating V, m, and R from these equations and rearranging:

$$T = T_i \frac{p}{p_i} \tag{2.13}$$

It is known that the ideal gas equation, and therefore eqn 2.13, are more accurate at low pressure. So the temperature is actually found by extrapolating to a pressure of zero by successively repeating the experiment with less and less gas in the thermometer:

$$T = T_i \lim_{p \to 0} \frac{p}{p_i} \tag{2.14}$$

This procedure is illustrated in Fig. 2.5. The constant-volume gas

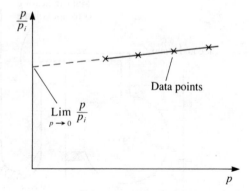

Fig. 2.5 The constant-volume gas thermometer: extrapolation to zero pressure

thermometer is not a practical method of measuring temperature, but it does provide an objective temperature scale.

A practical scale is the international temperature scale agreed in 1968. This scale defines the temperature at which various physical phenomena occur: for example, the freezing point of liquid gold is defined as 1337.58 K and the triple point[2] of oxygen as 54.361 K.

2.6 Problems

2.1 Chlorine has atomic number 17. It has a relative molar mass 35.4, and is known to exist in two isotopic forms—one with 18 neutrons and the other with 20 neutrons. Calculate the approximate relative abundance of these isotopes.

2.2 A room is 3 m by 4 m, and is 2.5 m high. Calculate the mass of air and the approximate number of oxygen and nitrogen molecules in the room. The pressure of the air is 1 bar and the temperature is 20°C.

2.3 Complete the following table of properties for ideal gases.

Substance	Quantity	Volume	Temperature	Pressure
helium	2 kg	? m^3	90 K	15 kPa
air	? kg	5 m^3	23°C	210 kPa
CO_2	3.1 kg-moles	60 m^3	?°C	400 kPa

[2]The triple point is the single temperature at which liquid, solid, and vapour can exist together.

2.4 Treating steam as an ideal gas, calculate the specific volume of steam under the following conditions. Compare your calculated answers with tabulated values from steam tables.

(a) saturated steam at 0.1 bar.

(b) saturated steam at 100 bar.

(c) superheated steam at 1 bar and 500°C.

(d) superheated steam at 200 bar and 500°C.

2.5 The temperature of a hot liquid is measured with a constant-volume gas thermometer (see Fig. 2.4). The reference temperature is melting ice and the pressure ratio p/p_i is measured at a number of values of p as given in the table below. Calculate the temperature of the liquid.

p (bar)	1.0	0.8	0.6	0.4	0.2
p/p_i	1.3049	1.3032	1.3008	1.2991	1.2969

2.6 The temperature of the atmosphere falls with height. If it is actually assumed that the temperature is constant, show that the variation of pressure with height is given by:

$$\ln \frac{p}{p_o} = -\frac{gz}{RT}$$

where p_o is the pressure at sea level where $z = 0$. From this equation calculate the pressure at heights of 1 km, 10 km, and 100 km. Assume that the temperature of the atmosphere is 300 K.

2.7 From the equation in question 2.6 (or otherwise) estimate the total number of molecules in the atmosphere of the earth. Take the radius of the earth to be 8000 km.

3

Real gases

3.1 Key points of this chapter

- Real gases behave ideally at low pressure and high temperature. (Section 3.2)

- Classical experiments of Andrews on carbon dioxide showed the real variation of volume with pressure at various temperatures. (Section 3.2)

- The critical point: its relation to the existence of the liquid phase and its definition in terms of the variation of volume with pressure. (Section 3.2)

- Various equations can be used to describe the behaviour of real gases: the Clausius equation, Van der Waals equation, and the virial equation. (Section 3.3)

- By using reduced properties: pressure, volume, and temperature made dimensionless using the values at the critical point, a generalized equation of state can be derived. This is the law of corresponding states. (Section 3.4)

3.2 Departures from ideal gas behaviour

In the last chapter we saw that gases are most ideal (that is they most closely obey the ideal gas equations) when the density of the gas is low. But what makes the gas behave non-ideally? One almost trivial answer is that real gases liquefy if the temperature becomes very low, whereas ideal gases do not change their fundamental behaviour however low the temperature. Some of the effects of the existence of solids and liquids as well as gases are described in Chapter 4.

However, even if they do not become liquid, real gases do not precisely obey the equation:

$$pv = RT \tag{3.1}$$

The molecules in a gas are in constant motion. It is possible using this picture of the gas—the kinetic theory of gases—to derive the ideal gas equation, eqn 3.1 (see Chapter 11). To perform this derivation it is necessary to make a number of basic assumptions about the gas molecules:

1. that the molecules are moving in random directions and are free to move in any direction,

2. that the pressure on the walls of a container arises as the molecules bounce off the wall—their momentum is changed and so a force must have been applied, and

3. that there are no forces on the molecules—in particular that there are no forces between the molecules.

In fact the molecules are not free to move in any direction: movement in some directions will cause a collision with another molecules. Also there is a force between molecules. Both these effects mean that eqn 3.1 is not actually obeyed. Before looking at some equations which have attempted to account for these real effects, it is useful to look at some classic experiments and results on real gas behaviour.

Andrews, in 1863, studied the relationship between pressure and volume of carbon dioxide at around room temperature. The gas was contained in a thick-walled glass capillary tube and was compressed by hydraulic pressure (see Fig. 3.1). When the volume was plotted against the pressure at various temperatures, results of the type shown in Fig. 3.2 were obtained. Note the following.

1. At high temperatures, the gas obeys the equation $pv = C$ reasonably well, but much less well as the temperature is reduced. At high temperatures there is no sign of the gas liquefying.

2. At low temperatures, as the pressure is increased, the gas liquefies. In the glass capillary tube the liquid CO_2 could clearly be seen. As the liquefaction is occurring the pressure remains constant. Once all the gas has liquefied, the liquid CO_2 is almost incompressible: as the pressure increases the volume hardly changes.

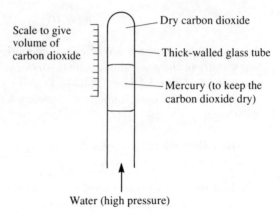

Fig. 3.1   Andrews' apparatus for studying the compression of carbon dioxide

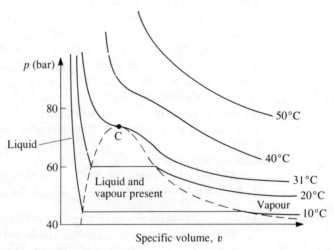

Fig. 3.2   Andrews' results for the compression of carbon dioxide

3. There is a definite temperature which separates these regions of high and low temperature: for CO_2 it is 31°C. This line of constant temperature just touches the region where both liquid and gas are present. At the point of touching, marked C in Fig. 3.2, mathematically:

$$\left(\frac{\partial p}{\partial v}\right)_T = 0 \tag{3.2}$$

and

$$\left(\frac{\partial^2 p}{\partial v^2}\right)_T = 0 \tag{3.3}$$

C is known as the critical point, and will be discussed further in the next chapter. At one end of the line of constant temperature passing through C (where the pressure is low), the CO_2 is certainly gaseous. At the other end (where the pressure is high) it certainly has the properties of liquid. But nowhere along the line is the gas seen to liquefy. At the point C the temperature is the critical temperature T_c, the pressure is the critical pressure p_c, and the specific volume is the critical specific volume v_c.

3.3 Equations which describe the behaviour of real gases

Many equations have been proposed to describe the behaviour of real gases: here three of them are discussed. These equations relate pressure, volume, and temperature: in general they are known as equations of state.

1. An equation which takes account of the fact that molecules cannot move to positions occupied by other molecules is the **Clausius equation**:

$$p(v - b) = RT \tag{3.4}$$

In eqn 3.4 b is a volume which is related to the actual volume of one kilogram of the molecules. At normal conditions the molecular volume is small, and so b is much smaller than v.

2. **Van der Waals equation** is an attempt to take molecular forces into account as well as the molecular volume. The effect of the molecular forces is to reduce the expected (ideal gas) pressure by an amount which is proportional to (density)^2. Using specific

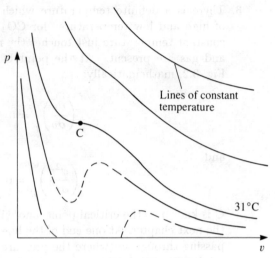

p

Lines of constant
temperature

C

31°C

v

Fig. 3.3   Van der Waals equation plotted for carbon dioxide

volume, v instead of density, p has to be replaced by $p + a/v^2$ to
give:

$$\left(p + \frac{a}{v^2}\right)(v - b) = RT \tag{3.5}$$

Both eqns 3.4 and 3.5 are written in terms of one kilogram of gas.
Figure 3.3 shows eqn 3.5 at various values of the temperature.
There is a considerable similarity between the theoretical results
of Fig. 3.3 and the experimental results of Andrews shown in
Fig. 3.2.

3. The **virial equation** is a more general equation relating pressure,
volume, and temperature, and is of the form:

$$pv = A + \frac{B}{v} + \frac{C}{v^2} + \frac{D}{v^3} + \cdots \tag{3.6}$$

The coefficients A, B, C, etc. are functions of temperature. Obviously if $A = RT$ and all the other coefficients are zero eqn 3.6
reduces to the ideal gas equation, eqn 3.1. By choosing other
values for the coefficients A, B, C, etc. the Clausius equation
and the Van der Waals equation (eqns 3.4 and 3.5) can be re-
constructed. Of course it is always true that $A = RT$. The
coefficients A, B, C, etc. are of decreasing importance in determining the actual behaviour of the gas. The advantage of
the virial equation over similar equations with a polynomial in

the pressure p on the right hand side is that eqn 3.6 has some theoretical justification.

Example 3.1

For carbon dioxide the Van der Waals constants are, in SI units per kg of gas:

$$a = 185.5 \quad \text{and} \quad b = 9.7 \times 10^{-4}$$

Carbon dioxide is contained in a vessel of volume 2 m³ at a temperature of 350 K. The pressure is 100 bar. Calculate the mass of gas using the ideal gas equation and Van der Waals equation.

Solution

The specific volume has to be calculated, and then $m = V/v$. For an ideal gas:

$$v = \frac{RT}{p} = \frac{(8314/44) \times 350}{100 \times 10^5} = 6.6 \times 10^{-3} \text{m}^3/\text{kg}$$

and the mass is $2/6.6 \times 10^{-3} = 302$ kg. For a Van der Waals gas the equation to be solved for v is:

$$\left(100 \times 10^5 + \frac{185.5}{v^2}\right)(v - 9.7 \times 10^{-4}) = (8314/44) \times 350$$

This is most easily solved by trial and error, and the solution for v is 4.2×10^{-3} m³/kg. The mass of gas is therefore $2/4.2 \times 10^{-3} = 476$ kg. In this case the two answers are significantly different. Here where the pressure is high and the temperature is relatively low, the gas is not behaving ideally.

Example 3.2

Repeat the previous example for a pressure of 1 bar and a temperature of 750 K.

Solution

For an ideal gas:

$$v = \frac{RT}{p} = \frac{(8314/44) \times 750}{1 \times 10^5} = 1.417 \text{m}^3/\text{kg}$$

and the mass is $2/1.417 = 1.411$ kg. For a Van der Waals gas the equation to be solved for v is:

$$\left(1 \times 10^5 + \frac{185.5}{v^2}\right)(v - 9.7 \times 10^{-4}) = (8314/44) \times 750$$

The solution of this equation is not detectably different from the solution for the ideal gas equation, and so again the mass of gas is 1.411 kg. Thus at these conditions of low pressure and high temperature, the gas is behaving ideally.

3.4   The law of corresponding states

It has been found that the real behaviour of many gases can be correlated by the law of corresponding states. If the temperature, pressure, and specific volume are expressed in terms of the reduced quantities T_r, p_r, and v_r defined by:

$$T_r = T/T_c \tag{3.7}$$

$$p_r = p/p_c \tag{3.8}$$

and

$$v_r = v/v_c \tag{3.9}$$

then behaviour of the gas can be expressed, in general terms, as:

$$p_r = f(T_r, v_r) \tag{3.10}$$

For example, it can be shown that Van der Waals equation (eqn 3.5) can be re-written in terms of the reduced properties as:

$$\left(p_r + \frac{3}{v_r^2}\right)\left(v_r - \frac{1}{3}\right) = \frac{8}{3}T_r \tag{3.11}$$

Equation 3.11 is obtained by using three equations true at the critical point—Van der Waals equation itself (eqn 3.5) eqns 3.2 and 3.3 to find the critical temperature, pressure, and specific volume in terms of the variables a, b, and R. Then a, b, and R are eliminated from eqn 3.5 to give the reduced property equation, eqn 3.11.

Another related way to express the behaviour of a real gas is to re-arrange the virial equation (eqn 3.6) into the form:

$$Z = \frac{pv}{RT} = 1 + \frac{B}{RTv} + \frac{C}{RTv^2} + \frac{D}{RTv^3} + \cdots \tag{3.12}$$

where Z is known as the gas compressibility. For an ideal gas Z should be equal to unity. Equation 3.12 can be re-arranged into the form:

$$Z = \frac{pv}{RT} = f(T_r, v_r) \tag{3.13}$$

Of course any two of the three reduced variables could be used on the right hand side of eqn 3.13. In practice the form of this equation which tends to be used is:

$$Z = \frac{pv}{RT} = f(T_r, p_r) \tag{3.14}$$

Forms of eqn 3.14 are often presented graphically, where Z is plotted against p_r at various values of T_r. An example is shown in Fig. 3.4.

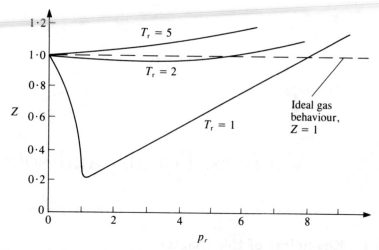

Fig. 3.4 Gas compressibility as a function of reduced temperature and pressure

From this figure it can be seen that the gas behaves ideally, that is Z is in the region of unity, when the pressure is low and the temperature is high.

3.5 Problems

3.1 In Section 3.3 values for the Van der Waals constants (a and b) are given for carbon dioxide. What are the units of these constants?

3.2 From Van der Waals equation (eqn 3.5) find equations for the critical temperature, pressure, and volume in terms of the constants a, b, and R in the equation. Hence find the value of the dimensionless group $p_c v_c / R T_c$.

3.3 From the answers to question 3.2 and the values of a and b for carbon dioxide in Section 3.3 calculate the values of the critical temperature, pressure, and volume.

3.4 Starting from Van der Waals equation (eqn 3.5) and from the results of question 3.2 prove eqn 3.11—Van der Waals equation written in terms of the reduced variables.

3.5 Using eqn 3.11 plot the compressibility Z against the reduced pressure as shown in Fig. 3.4 for a value of the reduced temperature of 1. [Hint: select values of the reduced volume, and then calculate Z and p_r. Values of v_r of 3, 1, and 0.5 are interesting.]

4

Vapours, liquids, and solids

4.1 Key points of this chapter

- The phase rule relates the number of variables that can be independently altered to the number of phases and the number of chemical species. (Section 4.2)

- In general two properties can be varied independently, but this is reduced to one for two-phase systems, and to zero for three-phase systems. (Section 4.2)

- In a vapour–liquid system the pressure and the temperature cannot be varied independently. The relationship between the two variables is the vapour pressure curve. (Section 4.3)

- The vapour pressure curve continues to high pressure, but stops at a definite point, the critical point. At this point the properties of the vapour and the liquid are identical. (Section 4.4)

- For a system like ice–water–steam the temperature, pressure, and specific volume can be plotted on a three-dimensional graph. This is the equilibrium diagram. (Section 4.5)

- The full equilibrium is rather complicated. Often just the projection on the pressure–volume plane is used. This is the p–v diagram. (Section 4.5)

- The properties of steam and water are usually tabulated in 'steam tables'. Here the values of the specific volume from the steam tables are examined. (Section 4.6)

4.2 The phase rule

The phase rule was first derived by Gibbs, one of the greatest contributors to our knowledge of thermodynamics. The phase rule will not be derived here—derivations can be found in many text books. The intention here is to look at a few consequences of the phase rule.

The phase rule can be expressed simply:

$$F + P = C + 2 \tag{4.1}$$

Here C is the number of chemical species present, P is the number of phases present (a phase meaning vapour, liquid, or solid), and F is the number of degrees of freedom. Here we are concerned only with systems where there is only one chemical species present, so C is equal to 1. The meaning of F, the number of degrees of freedom, is best explained by means of a number of particular cases.

1. One phase present. With just one phase P is equal to 1, and so from eqn 4.1 the number of degrees of freedom F is equal to 2. If a certain amount of the material (vapour, liquid, or solid) is present, then two variables can be chosen with complete freedom. This is best seen by reference to the ideal gas equation for a mass of gas, m:

$$pV = mRT \tag{4.2}$$

   Any two of the three variables p, V, and T can be chosen and then the consequence is that the third variable is automatically fixed. It is a general rule that for a single-phase system two variables are necessary and sufficient to define the system.

2. Two phases present—for example liquid and vapour. Here P is equal to 2, and so from the phase rule (eqn 4.1) F is equal to 1. There is thus a single degree of freedom. The implication of this is that only one variable can be chosen at will. So with a mixture of, for example, steam (water vapour is just another term for steam) and water only a single variable can be chosen and the others are fixed. It is impossible to fix both the temperature and the pressure: fixing the pressure automatically determines the temperature. Further details of this point are covered in the Section 4.3.

3. Three phases present—solid, liquid, and vapour all present. Now P is equal to 3, and so from the phase rule (eqn 4.1) there are no degrees of freedom F, so that no variables can be chosen at

Table 4.1 Consequences of the phase rule for a pure component system

Number of phases P	Example	Degrees of freedom F	Consequences
1	steam	2	two variables can be altered
2	steam & water	1	only one variable can be altered
3	steam, water, & ice	0	situation is fixed

will. If water, steam, and ice are all present together then the conditions are fixed. They can only exist together at a single pressure and temperature. This single point is the triple point, and for water occurs when the pressure is 0.00611 bar (611 Pa) and a temperature of 0.01°C (273.16 K). The fact that the triple point is a unique point makes it suitable as one of the points which defines the practical temperature scale.

This information from the phase rule is summarized in Table 4.1 for a pure component system, that is with $C = 1$.

4.3 The vapour pressure curve

As noted in Section 4.2 a system with two phases present has just one degree of freedom. One manifestation of this fact is the existence of the vapour pressure curve. The situation imagined is shown in Fig. 4.1. The closed box contains only the chemical substance under investigation, so if it is water there are only H_2O molecules and there is no air. Then as the box is heated and the temperature increases, the consequence of the phase rule is that, corresponding to every temperature, there is a pressure. The single degree of freedom is the temperature. Once this degree of freedom has been used up, the pressure is fixed. This pressure is the vapour pressure of the liquid.

The results from this procedure are shown in Fig. 4.2 where the vapour pressure is plotted against the temperature. Marked on Fig. 4.2 are regions corresponding to single-phase liquid and vapour systems. The liquid is sub-cooled liquid, that is liquid which is below its boiling temperature at the particular pressure. The vapour is superheated vapour—it has been heated above its boiling temperature. Along the curve the system comprises liquid and vapour in equilibrium. The phases are said to be saturated. For this reason the curve is sometimes called the saturation vapour pressure (S.V.P.) curve, and the temperature and pressure points along the curve are referred to as saturation temperatures and pressures.

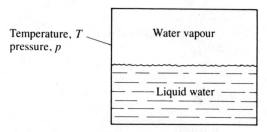

Temperature, T
pressure, p

Water vapour

Liquid water

Fig. 4.1 System for vapour pressure 'experiment'

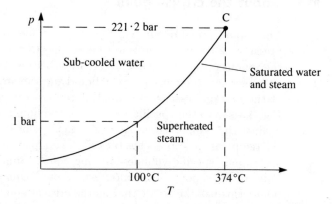

Fig. 4.2 Results from the vapour pressure 'experiment'

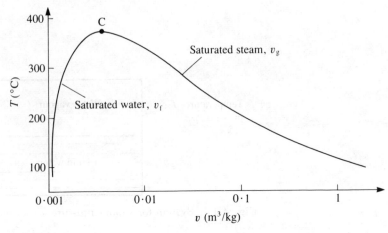

Fig. 4.3   Liquid and vapour specific volumes plotted against temperature for water

Figure 4.2 is actually plotted for the steam–water system. The vapour pressure curve passes through the normal boiling point of water: a vapour pressure of 1 bar at a temperature of 99.63°C.

4.4   More about the critical point

The vapour pressure curve in Fig. 4.2 continues to higher and higher pressures and temperatures until it reaches C, the critical point. Here the vapour pressure curve stops. It stops because at this point there is no longer any difference between liquid and vapour. All the properties of the two phases become equal at the critical point. For example Fig. 4.3 shows the specific volumes of the liquid and vapour plotted against the temperature. As the critical point C is approached and the temperature approaches the critical temperature T_c the liquid and the vapour specific volumes become equal. Similarly the refractive index of the vapour–liquid interface becomes equal to 1 at the critical point: this has the result that as the critical point is approached the meniscus enabling the interface to be seen gradually vanishes.

For water the conditions at the critical point are

$$p_c = 221.2 \text{ bar} \qquad T_c = 374.15°C \qquad v_c = 3.17 \times 10^{-3} \text{ m}^3/\text{kg}$$

Here the specific volume v_c is of course the specific volume of both

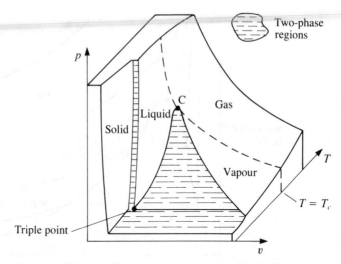

Fig. 4.4 Equilibrium diagram for a normal solid–liquid–vapour system

the phases: at the critical point there is no distinction between the liquid and the vapour phases.

4.5 Equilibrium diagrams for the vapour–liquid–solid system

Just as the diagrams in the last chapter illustrating Boyle's law ($pv = C$) and Charles' law ($p/T = C$) are two-dimensional projections of a three-dimensional surface ($pv = RT$), so here the vapour pressure curve (Fig. 4.2) and the curve of specific volumes against temperature (Fig. 4.3) are two-dimensional projections of a three-dimensional surface. Again the surface is a plot on a p, v, T diagram, and it shows the regions of existence of the various phases. The example taken here is that of a liquid which contracts on freezing—this is the normal behaviour but is not like water, which exhibits the anomalous behaviour that the solid phase (ice) is less dense the the liquid phase (water) with which it is in equilibrium. Ice floats in water, but in most substances the solid phase would sink when immersed in the liquid phase. The diagram is shown in Fig. 4.4.

The $p–v–T$ surface shown in Fig. 4.4 consists of a number of parts.

1. There is a curved surface representing each single phase region: solid, liquid, or vapour. If two of the co-ordinates of the surface are known (the two degrees of freedom) the third can be found from the diagram.

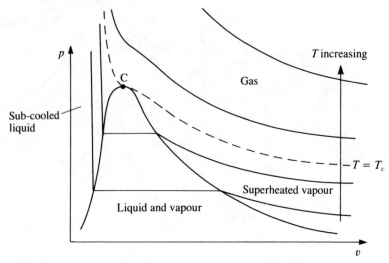

Fig. 4.5   *p–v* diagram showing the liquid and vapour phases

2. There is a surface for each combination of two phases: solid–liquid, solid–vapour, and liquid–vapour. These three surfaces necessarily touch at one single point: the triple point. This point is a unique point on the surface, as there are no degrees of freedom.

3. When lines of constant temperature are drawn, the constant temperature lines cross these three surfaces representing a two-phase region at constant pressure. In other words fixing the temperature fixes the pressure automatically, and there is only one degree of freedom.

The most useful version of this three-dimensional surface is a projection on the *p–v* plane showing a number of different lines of constant temperature (see Fig. 4.5). In this figure very low temperatures are not shown and thus the solid region is not shown. This figure is similar to Fig. 3.2 which showed the results produced by Andrews: here the phases are marked.

The convention used is that the gaseous phase is termed gas if it is above the critical temperature, and vapour if it is below the critical temperature. This accords with everyday usage: air is a 'gas', but water 'vapour' is used as a term for steam. Figure 4.5 will be used frequently to explain various processes occurring in the steam–water system.

Table 4.2   Saturated water and steam, to 100°C

t_s (°C)	p (bar)	v_f (dm³/kg)	v_g (dm³/kg)
0.01	0.006112	1.0002	206163
20	0.023366	1.0017	57838
50	0.123350	1.0121	12046
100	1.013250	1.0437	1673

Table 4.3   Saturated water and steam, to 221 bar

p (bar)	t_s (°C)	v_f (dm³/kg)	v_g (dm³/kg)
0.00611	0.01	1.0002	206163
0.10	45.83	1.0102	14673.7
1.00	99.63	1.0434	1693.7
10.00	179.88	1.1274	194.30
100.00	310.96	1.4526	18.04
221.20	374.15	3.1700	3.17

4.6   Steam tables and their use

The properties of steam and water are tabulated in what are commonly called 'steam tables', though they are as much about water as about steam. There are usually two main tables: the saturation table, and the superheat table.

1. The saturation table gives information about the properties of saturated water and saturated steam, that is along the saturation lines marked on Fig. 4.5. The saturation table is sometimes provided in terms of temperature, see Table 4.2, and sometimes in terms of pressure, see Table 4.3. Note that in thermodynamic tables, liquid phase is usually indicated by a subscript 'f', and the vapour phase a subscript 'g'.

2. The superheat table gives information about superheated steam, see the region marked in Fig. 4.5. Table 4.4 shows an extract from a typical superheat table. The format of this table appears rather unusual in that there are four horizontal lines each giving values of the superheated steam specific volume, v. In Chapters 7

Table 4.4 Specific volume of superheated steam, to 220 bar and 800°C

p (bar) (t_s (°C))		Temperature (°C)				
		t_s	100	200	400	800
1.00 (99.60)	v (dm³/kg)	1693.7	1695.5	2172.3	3102.5	4951.7
10 (179.9)	v (dm³/kg)	194.30		205.92	306.49	494.30
100 (311.0)	v (dm³/kg)	18.041			26.408	48.580
220 (373.7)	v (dm³/kg)	3.7347			8.2510	21.599

and 10 further versions of this table will be given where other properties, as well as the specific volume, are given.

Tables 4.2, 4.3, and 4.4 contain, in each case, only a few of the entries of the full table. Enough entries are given here to enable the use of the tables to be illustrated with a number of examples. Note that 1 dm³/kg = 10^{-3} m³/kg.

Example 4.1

Steam and water are present in a closed container. The temperature is 50°C, what is the pressure and the density of each phase? Compare the steam density with the ideal gas value.

Solution

Steam and water are both present, so the phases are saturated. The temperature is given, so the appropriate table is Table 4.2. From this table, the saturation pressure is 0.123 bar. Also from this table the specific volumes are (converting to SI units):

$$v_f = 1.0121 \times 10^{-3} \text{ m}^3/\text{kg} \quad v_g = 12046 \times 10^{-3} \text{ m}^3/\text{kg}$$

The density is the reciprocal of specific volume and so:

$$\rho_f = 988 \text{ kg/m}^3 \quad \rho_g = 0.083 \text{ kg/m}^3$$

The steam density can also be calculated from the ideal gas equation:

$$\rho_g = \frac{p}{RT} = \frac{0.123 \times 10^5}{(8314/18) \times 323} = 0.082 \text{ kg/m}^3$$

It is commonly found that the vapour density is well calculated by the ideal gas value unless the pressure is very high.

Example 4.2   A box of volume 10 m^3 contains 2000 kg of water and steam at 10 bar. Calculate the volume occupied by each phase, and the mass of each phase.

Solution   First it should be confirmed that there are both steam and water present. The average specific volume is $10/2000 = 0.005$ m^3/kg. This is between the water and steam specific volumes: so both phases actually are present. Now let the volume of water be V_f and the volume of steam be V_g, then the masses of each phase are V_f/v_f and V_g/v_g. So from our knowledge of the total volume and total mass:

$$V_f + V_g = 10$$

and

$$V_f/v_f + V_g/v_g = 2000$$

where from Table 4.3 $v_f = 0.0011274$ m^3/kg and $v_g = 0.19430$ m^3/kg. Solving these simultaneous equations for V_f and V_g gives the volume of the liquid as 2.21 m^3, and the volume of the steam as 7.79 m^3. The mass of the liquid is 1960 kg, the mass of the steam is 40 kg.

Example 4.3

Solution   Repeat example 2 with a total mass of 50 kg.

This time the average specific volume is $10/50 = 0.2$ m^3/kg. This is above the saturated steam value (0.19430 m^3/kg) at 10 bar, so the steam must be superheated. In the superheat table, Table 4.4, it can be seen that 0.2 m^3/kg lies between the saturation value and the value at 200°C which is 0.20592 m^3/kg. Linearly interpolating gives a temperature of 189.7°C. The pressure is now much higher than in the first example, and so the ideal gas equation is unlikely to work very well. Indeed if this method is used, the temperature would be calculated as 160°C, which is considerably in error.

4.7   Problems

4.1 Using the data for the critical point of water in Section 4.4, test the relation for $p_c v_c/RT_c$ derived in question 3.2 for a substance obeying Van der Waals equation.

4.2 Draw the analogue of Fig. 4.4 for water—a substance which expands on freezing.

4.3 Using steam tables complete the following table.

Mass	Volume	Temperature	Pressure	Dryness if saturated	Superheated or saturated
10 kg	10 m^3	?	1 bar	?	?
10 kg	31 m^3	?	1 bar	?	?
5 kg	10 m^3	600°C	?	?	?
100 kg	?	311°C	100 bar	?	?
?	100 m^3	?	1 bar	0.5	?

5

Systems, processes, and cycles: the language of thermodynamics

5.1 Key points of this chapter

This chapter is largely concerned with the definition of various thermodynamic concepts, from that of a system, to the ideas of reversibility and irreversibility which are important for the following chapters.

- Various types of system are defined, for example open systems, closed systems, and isolated systems. (Section 5.2)

- Reversible and irreversible processes. Examples of such processes. The causes of irreversibility. (Section 5.3)

- Thermodynamic cycles as ways of converting heat into work. (Section 5.4)

- Definition of thermal efficiency of a cycle. (Section 5.4)

- Definition and significance of the work ratio for a thermodynamic cycle. (Section 5.4)

5.2 Thermodynamic systems and properties

In thermodynamics the totality of the world is usually termed **the universe**, and is divided into two parts.

1. **The system** is the part of the universe in which we are particularly interested.

2. The rest of the universe is **the surroundings**.

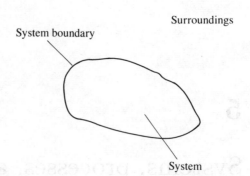

Fig. 5.1   The system, the surroundings, and the system boundary

Table 5.1   The various types of system

| System type | Boundary permeable to: | | |
	matter	heat	work
isolated	no	no	no
closed	no	yes	yes
open	yes	yes	yes

Separating the system from the surroundings is **the system boundary**, see Fig. 5.1. Systems can be of three types, depending on what can pass through the system boundary.

1. If matter, heat, and work can pass through the system boundary, then the system is **an open system**.

2. If heat and work can pass through the boundary, but matter cannot, then the system is **a closed system**.

3. If neither matter, not heat, nor work can pass through the boundary, then the system is **an isolated system**.

The various types of system are summarized in Table 5.1. The system might typically be a certain mass of gas, or a mixture of liquid and gas (such as water and steam). The condition of the system is identified by a number of **properties of state** such as the temperature, pressure, and volume. It will be noted later that heat and work are not properties of state like these other properties.

   The phase rule specifies how many properties of state are necessary to define the system (see Table 5.2). It is often stated that two

Table 5.2   Number of properties of state required to specify a system with a single pure component

Number of phases	Number of properties of state required
1	2
2	1
3	0

properties are necessary to specify a system, and this will be used frequently later, but it must be remembered that this is only true for a single–phase system.

Properties are sometimes divided into two types as follows.

1. **Extensive properties**: these depend on the mass of substance present, for example volume. The greater the mass of air, the greater will be the volume.

2. **Intensive properties**: these do not depend on the mass of substance present, for example pressure and temperature. Specific volume is defined as the volume per unit mass, and so must be an intensive property.

Extensive properties are usually given an upper case symbol, for example V for volume. Intensive properties are usually given a lower case symbol, so p for pressure, and v for specific volume. However, the use of T for the intensive property temperature is an exception to this rule.

5.3   Processes and reversibility

A process is a change in the system from one state to another. Hence the change in the pressure and temperature of a mass of ideal gas from (p_1, v_1) to (p_2, v_2) is a process. The state defined by (p_1, v_1) is the **initial state** of the system, and the state defined by (p_2, v_2) is the **final state** of the system.

Processes can be divided into two types as follows.

1. **Reversible processes**: these can be reversed so that both the system and the surroundings are returned to their original condition after the process and the reverse process have been carried out.

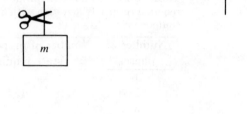

Initial   Final

Fig. 5.2   A mass is dropped onto a surface and comes to rest—an example of an irreversible process

2. **Irreversible processes**: these are ones in which this reversal cannot be carried out without leaving some change in the system or the surroundings.

The concept of reversibility is a very important one and is best illustrated by a number of examples.

1. If a mass is dropped onto a surface (see Fig. 5.2) the mass loses potential energy—a form of mechanical energy. This mechanical energy is lost as the mass comes to rest. A little of it is converted into sound energy which is absorbed by surrounding material and has the effect of heating the absorbing material very slightly. More is converted directly into heat by the action of friction in the material onto which the mass is dropped. This conversion of mechanical energy into heat energy is a sure sign of an irreversible process because as will be proved later, all this heat cannot be converted back into work.

2. If a ball-bearing rolls down a slope without friction the ball loses potential energy, as in the last example. However, this time the energy is converted into kinetic energy. This time it is possible to imagine that the process can be reversed, by causing the ball to roll up a similar slope (see Fig. 5.3). In theory and in the absence of friction the kinetic energy will be reconverted into potential energy, and the ball will come to rest at the original height. It is easy to see that any friction will cause the final

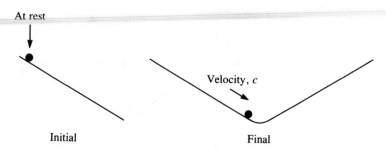

Fig. 5.3 A ball-bearing rolls down a slope—an example of an reversible process

Fig. 5.4 A cylinder containing compressed gas is confined by a movable piston

height to be less than the original height. In general, it is true that friction is a cause of irreversibility in real processes. The potential reversibility of this process occurs because one form of mechanical energy (potential energy) is converted into another form of mechanical energy (kinetic energy).

3. Figure 5.4 shows compressed gas in a cylinder. A force has to be applied to the piston to keep the gas compressed. Now if the force exerted by the gas on the piston is F_1 and the external force applied to the piston is F_2, and if F_1 is very slightly greater than F_2 then the piston will move slowly outwards and the gas will gradually expand. The pressure energy of the gas in the cylinder (a form of mechanical energy) is being converted into another form of mechanical energy as the force F_2 is pushed back—perhaps the piston is pushing against a spring. This slow expansion and inter-conversion of mechanical energy is another example of a reversible process. If now F_2 is very slightly greater than F_1 the gas in the cylinder will be slowly recompressed. Again friction, this time between the piston and the cylinder, will tend to cause irreversibility.

4. Alternatively, the restraining force on the piston in Fig. 5.4 could

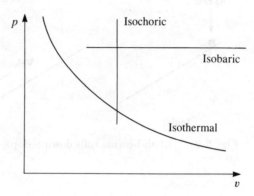

Fig. 5.5   Various types of process on a p–v diagram

be suddenly removed, and the gas in the cylinder allowed to expand quickly. The pressure energy of the gas has decreased, but what has happened to the energy? It has been dissipated in friction at the walls of the cylinder, and in friction between rapidly moving adjacent streams of gas by means of viscous dissipation. The mechanical pressure energy has been turned into heat energy, and so the process is an irreversible one. Expansion of a gas in a cylinder, as in Fig. 5.4, can only be reversible if the expansion is **fully resisted**, in other words if the forces acting on the cylinder are opposite and almost equal.

Processes can be described in other ways: if a process is carried out at constant pressure it is said to be **isobaric**, if it at constant temperature it is said to be **isothermal**, and at constant volume **isochoric**. **Diabatic processes** may exchange heat with the surroundings, but **adiabatic processes** are thermally insulated from the surroundings.

Processes can be illustrated as a series of points on a plot of two properties of state. A plot of pressure against specific volume, a p–v diagram, is often used for this purpose (see Fig. 5.5) where isobaric, isochoric, and (for an ideal gas) isothermal processes are marked. This latter process has the equation for an ideal gas:

$$pv = C \tag{5.1}$$

5.4 Thermodynamic cycles

A **thermodynamic cycle** is a linked series of processes such that the outlet of the final process is the same state as the input of the first process. It will be seen later that when a thermodynamic cycle is used to convert heat into work, heat must necessarily be rejected from the cycle. Figure 5.6 shows the block diagram of a generalized cycle. At various points heat and work are input and output. Two important quantities are the **thermal efficiency** of the cycle, η_{th}, and the **work ratio**, r_w defined by:

$$\eta_{th} = \frac{\text{net work out}}{\text{heat in}} = \frac{W_{out} - W_{in}}{Q_{in}} \tag{5.2}$$

$$r_w = \frac{\text{net work out}}{\text{work out}} = \frac{W_{out} - W_{in}}{W_{out}} \tag{5.3}$$

The thermal efficiency is a measure of how much of the input heat is converted into useful work, and makes the assumption that the heat rejected in the cycle is valueless. This is often, but not always, true. In an internal combustion engine the input heat is the heat liberated by the burning of the fuel. Of course a good cycle has a high thermal efficiency. A consequence of the first law of thermodynamics is that the maximum value possible for the thermal efficiency is 1.

The work ratio measures the relative sizes of the work out and the work in terms in Fig. 5.6. If they are of comparable size, giving a low work ratio, then there are two undesirable consequences:

1. For a given value of the net work out, individual terms W_{out} and W_{in} will be large. Thus large machines will be required to handle these large work transfers.

2. If the machines forming the cycle are not as mechanically efficient as expected, then W_{out} will be smaller than expected, and W_{in} will be larger than expected. This has the effect of drastically reducing the net work output below its expected value.

Hence a high work ratio is a desirable aim. The maximum value of the work ratio is 1.

5.5 Problems

5.1 Two thermodynamic cycles each have a thermal efficiency of 40%. One cycle has a value of $W_{in} = 0$ and the other has $W_{in} = \frac{2}{3}W_{out}$. If, for each cycle $Q_{in} = 100$, calculate for each cycle the values of Q_{out}, W_{in}, W_{out}, and the work ratio.

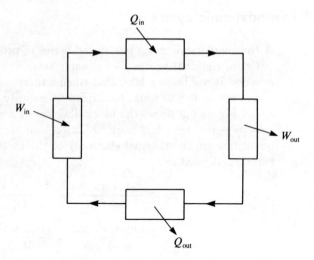

Fig. 5.6 A block diagram of a generalized thermodynamic cycle showing transfers of heat and work in and out of the cycle

5.2 For both of the cycles in question 5.1 Q_{in} is unaltered, but W_{out} is reduced by 10% and W_{in} is increased by 10%. Calculate the thermal efficiencies of the two cycles. Comment on the results.

6

The first law of thermodynamics for closed systems

6.1 Key points of this chapter

- The first law of thermodynamics, and the sign conventions used here for work and heat transfers. (Section 6.2)

- Internal energy, unlike heat and work, is a property of state (Section 6.3)

- Internal energy, in general, is a function of two other variables, for example temperature and pressure. (Section 6.3)

- The equation for the work in a reversible, closed-system process. (Section 6.4)

- Internal energy is related to the specific heat at constant volume. (Section 6.5)

- For an ideal gas, the internal energy is a function of the temperature only. (Section 6.5)

- Various examples are given for the use of the first law in closed systems with ideal gases. (Section 6.6)

- Examples with steam and water are in principle no more difficult, but are much more tedious requiring the use of tabulated data. (Section 6.7)

- The heat and work transfers in a reversible polytropic process with an ideal gas are calculated. (Section 6.8)

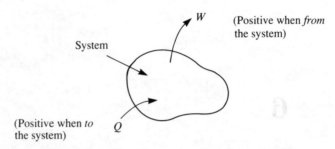

Fig. 6.1   Sign convention for heat and work

- Four polytropic processes (two isothermals and two adiabatics) can be linked to form a cycle to convert heat into work. (Section 6.8)

6.2   Statement of the first law

The first law can be stated simply, around a cycle:

$$\sum \delta Q = \sum \delta W \qquad (6.1)$$

where Q is the heat transfer to the system and W is the work done by the system. In eqn 6.1 the symbol $\sum$ means summation round a cycle. The sign convention for Q and W is stated above, and is summarized in Fig. 6.1. This sign convention is not universally used, so care has to be taken.

6.3   Internal energy

The existence of a property of state, the internal energy, is a consequence of the first law. Consider a diagram of two properties of state, for example pressure and volume, plotted against each other (see Fig. 6.2). Two states, 1 and 2, are marked on this diagram, and then a path A is drawn from 1 to 2. Two alternative return paths from 2 to 1 are shown: these are B and C. Two different cycles can now be made up, each going from point 1 to point 2, and then returning to point 1. One cycle is made up from path A and path B: the other from path A and path C. Applying eqn 6.1 to both these cycles, we obtain:

$$Q_A + Q_B = W_A + W_B \qquad (6.2)$$

and

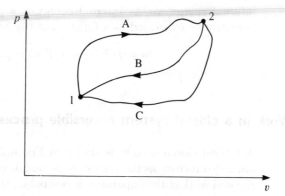

Fig. 6.2   Two cycles shown on a p–v diagram

$$Q_A + Q_C = W_A + W_C \tag{6.3}$$

Subtracting these equation eliminates all reference to path A, resulting in:

$$Q_B - W_B = Q_C - W_C = (Q - W)_{2 \to 1} \tag{6.4}$$

where $(Q - W)_{2 \to 1}$ indicates the value of $(Q - W)$ evaluated along any path going from point 2 to point 1. The conclusion therefore is that $(Q - W)$ is only dependent upon the end points (1 and 2), and does not depend at all on the path. Hence $(Q - W)$ is a property of state which is known as the internal energy. Note that Q and W are not themselves properties of state, but $(Q - W)$ is.

Internal energy is normally given the symbol U, and has units of energy (J). More frequently used is the specific internal energy: symbol u, units (J/kg).

Using internal energy, the closed system version of the first law can be written more usefully as:

$$Q - W = \Delta U = U_2 - U_1 \tag{6.5}$$

where U_1 is the initial internal energy, and U_2 is the final internal energy, and its differential form:

$$dQ - dW = dU \tag{6.6}$$

If Q and W are the heat and work transfers per unit mass of substance, then U in eqns 6.5 and 6.6 would be replaced by the specific internal energy u. As internal energy is a property of state it can normally be expressed as a function of two other state variables, so:

$$u = f(T, v) = f(T, p) = f(p, v) \qquad (6.7)$$

6.4   Work in a closed system reversible process

A typical closed system is shown in Fig. 6.3. If the cylinder moves out a distance δx as the gas expands, and also if the expansion is fully resisted so that the expansion is reversible, then the force restraining the motion of the piston is equal to the force exerted by the gas, F. If the pressure of the gas is p, and the cross sectional area of the cylinder is A, then:

$$F = pA \qquad (6.8)$$

The work done, δW, by the force F when it moves a distance δx is:

$$\delta W = F\,\delta x \qquad (6.9)$$

or using pressure in place of the force in eqn 6.9:

$$\delta W = pA\,\delta x \qquad (6.10)$$

Finally putting $A\delta x$ as δV, the volume swept out by the piston, we find that:

$$\delta W = p\,\delta V \qquad (6.11)$$

Note here that the actual volume of the gas V is used to give the work done by the entire mass of gas. If v is used in eqn 6.11, then the work would be the work done per kilogram of gas. Equation 6.11 is for a differential change. The equation can also be integrated to give:

$$W = \int_i^f p\,dv \qquad (6.12)$$

Here i represents the initial state in the process, and f represents the final state in the process.

It is important to realise when eqns 6.11 and 6.12 can be used. The conditions are that the system is closed, and that the process is reversible. The equations are true for any substance: solid, liquid, gas or any mixture of phases—certainly not just for an ideal gas. W is the work done **by** the system, and the work can be evaluated from eqns 6.11 and 6.12 as long as the relationship between pressure and volume is known.

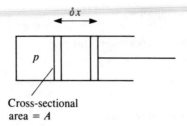

Cross-sectional
area = A

Fig. 6.3 A typical closed system: gas in cylinder with a piston

6.5 Internal energy and specific heat

Consider a form of eqn 6.7 applied to any substance:

$$u = f(T, v) \tag{6.13}$$

This equation can be differentiated to give:

$$du = \left(\frac{\partial u}{\partial T}\right)_v dT + \left(\frac{\partial u}{\partial v}\right)_T dv \tag{6.14}$$

Also, from the first law of thermodynamics du can also be written as:

$$du = dQ - dW \tag{6.15}$$

Now eqns 6.14 and 6.15 can be combined for the particular case of constant volume, or $dv = 0$ and therefore $dW = 0$ giving:

$$dQ = \left[\left(\frac{\partial u}{\partial T}\right)_v dT\right]_{\text{constant } v} \tag{6.16}$$

or, re-arranging this equation:

$$\left(\frac{\partial Q}{\partial T}\right)_v = \left(\frac{\partial u}{\partial T}\right)_v \tag{6.17}$$

The left-hand side of eqn 6.17 must be a specific heat as it is the amount of heat necessary to raise the temperature of one kilogram of substance by one degree, and so has SI units of J/kg K. The whole process is carried out at constant volume, so it is the specific heat at constant volume, c_v. Hence:

$$c_v = \left(\frac{\partial u}{\partial T}\right)_v \tag{6.18}$$

This is a perfectly general relationship between c_v and u.

A consequence of the ideal gas equation will be seen in Sections 11.3 and 18.3 to be that for an ideal gas, the internal energy depends only on the temperature and so:

$$c_v = \frac{du}{dT} \tag{6.19}$$

and therefore that:

$$\left(\frac{\partial u}{\partial v}\right)_T = 0 \tag{6.20}$$

Equations 6.19 and 6.20 are only true for ideal gases. Equation 6.19 can be integrated if the dependence of the specific heat with temperature is known. A common assumption is the specific heat is independent of temperature, so

$$u_2 - u_1 = c_v(T_2 - T_1) \tag{6.21}$$

An abbreviated form of this equation is often used:

$$u = c_v T \tag{6.22}$$

Here the constant of integration has been omitted. This does not matter as only differences in internal energy have any real meaning. Tables of internal energy always assign a zero internal energy at a more or less arbitrary temperature.

6.6   Examples of the use of the first law: ideal gases

The use of the first law for closed systems will be illustrated with a number of examples: in this section with ideal gases, and in Section 6.7 for steam–water systems. The common assumption is made that gases like air at low or moderate pressures behave like ideal gases. The value of c_v for air at room temperature is 718 J/kg K.

Example 6.1   A thermally insulated rigid box contains two compartments of equal volume. Initially one contains air at 5 bar and 25°C (298 K), and the other is evacuated (see Fig. 6.4). The dividing partition is then removed. Calculate the final air temperature and pressure.

Solution   If the system boundary is taken as the whole box (both compartments), then the system is isolated. No heat can cross the boundary as the box is insulated, and the system can do no work as the volume does not change. So from the first law, the internal energy must be

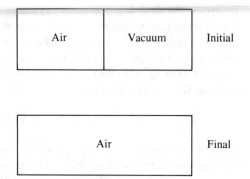

Fig. 6.4   Ideal gas: example 6.1—initial and final states

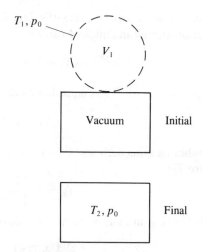

Fig. 6.5   Ideal gas: example 6.2—initial and final states

constant. For an ideal gas the internal energy is a function of temperature only. If the internal energy does not change, the temperature must remain constant. So the final temperature is 298 K. As the volume doubles at constant temperature, from the ideal gas equation the pressure must halve, so the final pressure is 2.5 bar. This process of sudden expansion is an irreversible process.

Example 6.2   A thermally insulated, rigid, evacuated box is suddenly punctured and air enters from the atmosphere (at 298 K), see Fig. 6.5. Calculate the final temperature of the air in the box.

Solution   To apply the first law for closed systems to this problem, a system boundary across which no mass flows must be drawn. This boundary can be imagined as a rubber balloon, as shown in Fig. 6.5, which collapses as the air enters the box. As in the previous problem it is assumed that $Q = 0$, but this time $W \neq 0$. The atmosphere does work on the system, and so:

$$W = -p_0 V_1 \tag{6.23}$$

where p_0 is the atmospheric pressure, and V_1 is the volume of air entering the box (see Fig. 6.5). The minus sign is because work is done on the system. So for this problem the first law becomes:

$$p_0 V_1 = U_2 - U_1 \tag{6.24}$$

If the mass of gas entering the box is m, then the terms involving internal energy in eqn 6.24 can be written in the form $mc_v T$, so that:

$$p_0 V_1 = mc_v (T_2 - T_1) \tag{6.25}$$

Applying the ideal gas equation to the air in its original state gives:

$$p_0 V_1 = mRT_1 \tag{6.26}$$

Combining eqns 6.25 and 6.26 gives an equation for the final temperature T_2:

$$T_2 = T_1 \left[\frac{R + c_v}{c_v} \right] \tag{6.27}$$

Substituting numerical values in this equation gives:

$$T_2 = 298 \left[\frac{\frac{8314}{29} + 718}{718} \right] = 417 \text{ K} = 144°\text{C}$$

This derivation ignores the heat capacity of the walls of the box. In any attempt to measure the temperature in this kind of experiment, this heat capacity term actually dominates, and so the temperature measured would only be slightly above ambient temperature. The derivation also assumes that the air enters the box slowly enough to be approximately reversible so that eqn 6.23 can be used.

This example also provides an illustration of the value of a p–V diagram for the process (see Fig. 6.6). The process is from point 1 to point 2. The area under the curve is the work, since $W = \int p dV$. This area is easily seen to be $-p_0 V_1$. The next example shows the use of a p–V diagram even more clearly.

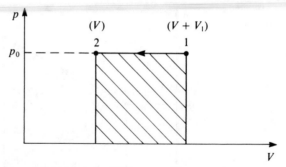

Fig. 6.6 *p–V* diagram for example 6.2

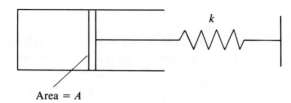

Fig. 6.7 Ideal gas: example 6.3—piston and cylinder

Example 6.3

Figure 6.7 shows a piston and cylinder arrangement where any expansion of the gas in the cylinder compresses a spring. The cross-sectional area of the cylinder is A, and the spring constant (the force per unit compression) is k. Initially the air is at atmospheric pressure p_0 and temperature T_0, and so there is no force in the spring. The cylinder is now heated slowly. Calculate the heat supply necessary to reach a specified pressure p_2.

Solution

The steps in the solution are:

1. Calculate the relationship between the pressure, p, in the cylinder at any point in the expansion, and the volume of the cylinder, V, by means of a force balance:

$$\frac{V - V_1}{A} k = (p - p_0)A \tag{6.28}$$

In this equation the left-hand side is the piston displacement multiplied by the spring constant, and the right hand side is the

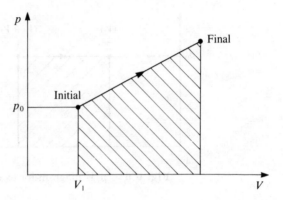

Fig. 6.8   Ideal gas—example 6.3: p–V diagram

pressure difference across the cylinder multiplied by the area. Equation 6.28 can be rearranged to give:

$$p = p_0 + (V - V_1)\frac{k}{A^2} \tag{6.29}$$

This equation is plotted in Fig. 6.8.

2. As the cylinder is heated slowly, the gas expands reversibly, and the work done by the gas is the shaded area in Fig. 6.8.

3. The mass of air in the cylinder, m, can be calculated from an ideal gas equation applied to the original conditions:

$$p_0 V_1 = mRT_1 \tag{6.30}$$

where T_1 is the initial temperature of the air in the cylinder.

4. The final temperature can be found from the ideal gas equation at the final conditions:

$$p_2 V_2 = mRT_2 \tag{6.31}$$

5. The change in the internal energy is calculated from:

$$U_2 - U_1 = mc_v(T_2 - T_1) \tag{6.32}$$

6. Finally as the work and the change in internal energy are known, the heat transfer can be calculated from the first law:

$$Q - W = U_2 - U_1 \tag{6.33}$$

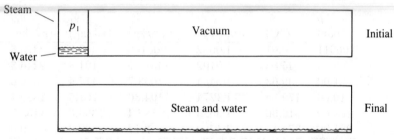

Fig. 6.9   Example 6.4: steam–water

This is a common type of problem. Because the relationship between p and V is known the work can be calculated. Then the internal energy change can be calculated, and finally the heat transfer can be found from the first law.

6.7   Example of the use of the first law: steam–water systems

In principle steam–water problems are similar to ideal gas ones. They are in practice much more complex because the properties of the steam and water have to be obtained from tables. Steam tables give values of the internal energy of steam and water (see for example Table 6.1 which shows part of the full saturation table). In this table:

u_f =   internal energy of saturated water (kJ/kg);
u_g =   internal energy of saturated steam (kJ/kg); and
u_{fg} = $u_g - u_f$ (kJ/kg).

An example illustrating first law calculations with steam–water is given here.

Example 6.4   Figure 6.9 shows a problem rather similar to the first ideal gas problem. A rigid, thermally insulated vessel is divided into two compartments. One compartment (volume 9 m^3) is evacuated, the other (volume 1 m^3) contains a mixture of saturated water and steam at 10 bar. The masses of each phase present is defined by **the dryness** or **the quality**, x:

$$x = \frac{\text{mass of steam}}{\text{total mass of steam and water}} \qquad (6.34)$$

Here the quality is 0.5.

Solution   As in the corresponding ideal gas problem, the total internal energy in the total volume (10 m^3) is constant. As the mass contained in the

Table 6.1   Values of v and u for saturated water and steam, to 221 bar

p (bar)	t_s (°C)	v_f (dm^3/kg)	v_g (dm^3/kg)	u_f (kJ/kg)	u_{fg} (kJ/kg)	u_g (kJ/kg)
0.00611	0.01	1.0002	206163	0.0	2375.6	2375.6
0.10	45.83	1.0102	14673.7	191.8	2246.3	2438.1
1.00	99.63	1.0434	1693.7	417.4	2088.6	2506.0
10.00	179.88	1.1274	194.30	761.5	1820.4	2581.9
100.00	310.96	1.4526	18.04	1393.6	1153.7	2547.3
221.20	374.15	3.1700	3.17	2037.3	0.0	2037.3

vessel is constant, the specific internal energy must also be constant. The initial state can be defined by the initial specific volume and the initial internal energy, v_1 and u_1. In terms of the quality, x, these are given by:

$$v_1 = (1 - x)v_{f1} + xv_{g1} \tag{6.35}$$

$$u_1 = (1 - x)u_{f1} + xu_{g1} \tag{6.36}$$

In these equations the first term represents the internal energy of the liquid and the second term that of the vapour. Substituting values from Table 6.1 gives:

$$v_1 = 0.0977 \text{ m}^3/\text{kg} \quad \text{and} \quad u_1 = 1671.7 \text{ kJ/kg}$$

The total volume increases by a factor of 10, but the final internal energy is unchanged. Hence the final state is defined by:

$$v_2 = 0.977 \text{ m}^3/\text{kg} \quad \text{and} \quad u_2 = 1671.7 \text{ kJ/kg}$$

The procedure to solve the problem is then one of trial and error. First a final pressure p_2 is guessed. From v_2, v_{f2}, and v_{g2} the value of the new quality x_2 can be calculated from an equation like eqn 6.35. Then using this value of x_2, and also the values of u_{f2}, and u_{g2} from the analogous equation to eqn 6.36, u_2 can be calculated and its value checked. The final pressure is then adjusted until a value of 1671.7 kJ/kg is obtained. In this case the final solution is:

$$p_2 = 1.044 \text{ bar} \quad \text{and} \quad x_2 = 0.599$$

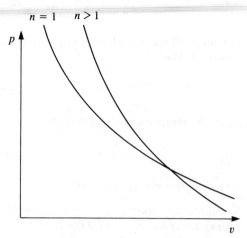

Fig. 6.10   Polytropic processes: the effect of the exponent n

6.8   Work and heat transfers in polytropic processes in gases

A polytropic process is one which follows an equation of the type:

$$pv^n = C \qquad (6.37)$$

when the exponent n is equal to 1, eqn 6.37 corresponds to Boyle's law. Real processes in gases usually correspond to values of n greater than 1. Fig. 6.10 shows the p–v diagram for a range of values of n. If the process is carried out reversibly, then the work per unit mass of gas can be calculated from $W = \int p\,dv$. This is true whether or not the gas is ideal. The results of this integration between an initial state 1 and a final state 2 are:

$$\text{for } n = 1: \qquad W = p_1 v_1 \ln \frac{p_1}{p_2} \qquad (6.38)$$

$$\text{for } n \neq 1: \qquad W = \frac{p_2 v_2 - p_1 v_1}{1 - n} \qquad (6.39)$$

If now additionally the gas is assumed to be ideal, then eqns 6.38 and 6.39 become respectively:

$$\text{for } n = 1: \qquad W = RT_1 \ln \frac{p_1}{p_2} \qquad (6.40)$$

$$\text{for } n \neq 1: \quad W = \frac{R(T_2 - T_1)}{1 - n} \quad (6.41)$$

For an ideal gas the change of internal energy $\Delta u = u_2 - u_1$ in the process is given by:

$$\text{for } n = 1: \quad \Delta u = 0 \quad (6.42)$$

since the temperature, and therefore the internal energy, is constant.

$$\text{for } n \neq 1: \quad \Delta u = c_v(T_2 - T_1) \quad (6.43)$$

The heat transfer Q can then be calculated from $Q - W = \Delta u$.

Example 6.5   Calculate the values of W, Δu, and Q for an ideal gas with the properties of air, $R = 287$ J/kg K and $c_v = 718$ J/kg K, as it expands from 2 bar, $25°$C to 1 bar with a polytropic exponent n of 1.1.

Solution   The gas obeys the equations $pv/T = C$ and $pv^n = C_1$. Eliminating v gives an equation relating the temperature and the pressure:

$$T_2 = T_1 \left(\frac{p_2}{p_1} \right)^{\frac{n-1}{n}} \quad (6.44)$$

From eqn 6.44:

$$T_2 = 298 \left(\frac{1}{2} \right)^{\frac{1.1-1}{1.1}} = 280 \text{ K}$$

Then from eqns 6.41 and 6.43:

$$W = \frac{287(280 - 298)}{1 - 1.1} = +52000 \text{ J/kg}$$

$$\Delta u = 718(280 - 298) = -13000 \text{ J/kg}$$

Finally from $Q - W = \Delta u$ the heat transfer is $+39000$ J/kg. So the temperature and therefore the internal energy decrease, work is done by the system, and there is heat transfer to the system.

Example 6.6   Repeat example 6.5 with values of the polytropic exponent n increasing from 1.0 to 1.5 in steps of 0.1.

Solution   The detailed calculation is not given here, but the results are given in Table 6.2. Clearly values of the exponent n of 1.0 and 1.4 are special for this gas (which has the properties of air).

Table 6.2 Values of work and heat transfers
for various polytropic processes—example 6.6

n	T_2 (K)	W (kJ/kg)	Δu (kJ/kg)	Q (kJ/kg)
1.0	298	+59	0	+59
1.1	280	+52	-13	+39
1.2	265	+47	-24	+23
1.3	254	+42	-32	+10
1.4	244	+38	-38	0
1.5	237	+35	-44	-9

1. When $n = 1.0$ the process, as expected, is isothermal or constant temperature. It should also be noted that the work done by the system is a maximum in this isothermal process. However, an isothermal expansion is difficult to achieve in practice because large amounts of heat have to be transferred to the system.

2. When $n = 1.4$ the process is adiabatic, that is there is no heat transfer with the surroundings. Processes which are approximately reversible and adiabatic are much easier to achieve in practice than reversible, isothermal processes.

It is also worth noting that when a gas expands it cools down considerably. Conversely, compression tends to raise the temperature of a gas.

6.9 A cycle to convert heat into work

Figure 6.11 shows a cycle made up from two isothermal processes and two adiabatic processes represented on a p–v diagram. If the cycle is in the direction of the arrows marked on the diagram, then the net result is to produce a net positive work output, and a net positive heat input. Thus heat is converted into work. There are two parts of the cycle involving heat transfer. From C to D, heat is taken into the cycle at a high temperature. From A to B, heat is rejected from the cycle at a low temperature. This is a general result: in cycles which convert heat into work, high temperature heat is taken into the cycle and low temperature heat is rejected. Table 6.3 shows the sign of the work and heat transfers in each of the individual processes. If all the individual processes are reversible, the work transfer per unit mass of

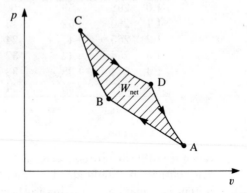

Fig. 6.11 Cycle: two isothermals, and two adiabatics

Table 6.3 Work and heat transfers in a cycle of two isothermals and two adiabatics (see Fig. 6.11)

Process	Process type	Work done by process	Heat transfer to process
A → B	isothermal	negative	negative
B → C	adiabatic	negative	zero
C → D	isothermal	positive	positive
D → A	adiabatic	positive	zero
overall cycle		positive	positive

gas during each process is given by $\int p\,dv$. For the whole cycle, the net work transfer, W_{net} is given by:

$$W_{net} = \oint p\,dv \qquad (6.45)$$

and W_{net} is equal to the area enclosed by the cycle on the p–v diagram.

The direction of the processes in the cycle in Fig. 6.11 is clockwise on the p–v diagram. Such clockwise cycles convert heat into work. Anti-clockwise cycles have the net result of converting work into heat.

6.10 Problems

6.1 A thermally insulated, rigid vessel is divided into two equal compartments. Each contains air which here behaves ideally. One compartment is at 2 bar and 300 K, the other is at 5 bar and 400 K. The partition between the compartments is ruptured. Calculate the final temperature and pressure of the air.

6.2 A cylinder of internal diameter 75 mm is closed at one end and sealed by a frictionless piston loaded by a spring of stiffness 20 kN/m. The cylinder initially contains 3×10^{-4} m^3 of air at 3 bar. Heat is added to the air such that the piston moves outwards by 30 mm. Calculate the work done by the air, the change in internal energy of the air and the heat transferred to the air.

6.3 1 kg-mole of air at 10 bar and 200°C expands reversibly along the path $pV^{1.2}$ =constant until the pressure falls to 1 bar. Calculate the work done by the air, the change in internal energy, and the heat transfer to the air. Assume the air is an ideal gas.

6.4 Repeat question 6.3 with 1 kg-mole of steam at the same conditions.

6.5 The heat transfer in question 6.3 is positive, but in question 6.4 is negative. Explain how this difference arises.

6.6 A sealed, thermally insulated tank of volume 2 m^3 has a safe working pressure of 4 bar. At 20°C 10% of the volume is occupied by water, the remainder by water vapour. Calculate how much heat can be added without exceeding the safe working pressure.

6.7 Repeat question 6.6 for a tank with a safe working pressure of 200 bar. At the pressure of 200 bar, what is the temperature?

6.8 A thermally insulated, rigid vessel is divided into two equal compartments. One contains steam at 100 bar and 400°C, and the other is evacuated. The partition is removed. Calculate the resulting pressure and temperature.

6.9 The pressure inside a rubber balloon is greater than atmospheric pressure, and the pressure above atmospheric pressure is proportional to the radius of the balloon. The initial radius is 0.1 m and the pressure inside the balloon is 1.1 bar. The balloon is inflated slowly, so that the temperature remains constant, until the radius is 0.2 m. Calculate the work done against the atmosphere and in stretching the rubber.

7

The first law of thermodynamics for steady-flow systems: the steady-flow energy equation

7.1 Key points of this chapter

- In a flow system when considering energy and the first law of thermodynamics, it is necessary to consider potential energy and kinetic energy as well as internal energy. (Section 7.2)

- A turbine extracts work from a high-pressure fluid as the pressure falls. (Section 7.2)

- The steady-flow energy equation is the version of the first law of thermodynamics which is applicable to steady flows. (Section 7.3)

- The property, enthalpy which appears in the steady-flow energy equation, is a property of state and is in general a function of two other properties of state. (Section 7.3)

- Enthalpy is related to the specific heat at constant pressure. (Section 7.4)

- Just as the work in a reversible closed system process is simply related to pressure and volume, so is the steady-flow reversible work. Both types of work are represented by different areas on a pressure–volume diagram. (Section 7.5)

- A number of simple examples of the use of the steady-flow energy equation with an ideal gas are given. (Section 7.6)

- Examples with steam and water are, in principle, no more difficult. All the properties have to be obtained from charts and
  tables, which makes the calculations more tedious. (Section 7.7)

- An example of a steam–water cycle that can be analysed with the
  steady-flow energy equation is the Rankine cycle. The efficiency
  and work ratio of a typical cycle are calculated. (Section 7.7)

7.2   More about work and internal energy

The work transfer discussed in the last chapter, where $W = \int p\,dv$ was
used for a closed-system reversible process, is also known as pv work.
It was assumed that if there was no change in volume, there could be
no work done. However, consider Fig. 7.1. Here gas flows from vessel
1 where the pressure is high, to vessel 2 where the pressure is low. As
the gas passes from one vessel to another it turns a series of blades in
a **turbine**. The rotating shaft S can be used to do external work. The
system of the two inter-connected vessels is certainly doing work on
the surroundings although there is no pv work. This rotating shaft
work is **shaft work** and is commonly found in steady-flow systems.

So far it has not been necessary to know the form of the internal
energy of a substance. In fact the internal energy is represented by the
internal energy of the molecules. The molecules can move randomly
and have kinetic energy of motion, they can rotate and have kinetic
energy of rotation, and they can internally vibrate and have both
kinetic energy and potential energy of vibration (for further details
see Chapter 11).

The first law for closed systems:

$$Q - W = \Delta u \tag{7.1}$$

is really a version of the law of conservation of energy. The energy
transfers during the process, by way of Q and W, are equal to the
difference in the energy content of the system before and after the
process, Δu. So really the difference in the energy content of the
system ought to include all kinds of energy. A more general statement
of eqn 7.1 is therefore:

$$Q - W = \Delta e \tag{7.2}$$

where e is the total energy per unit mass of the system—internal
energy, potential energy, and kinetic energy due to gross motion, so:

$$e = u + gz + \frac{c^2}{2} \tag{7.3}$$

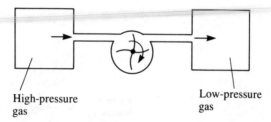

Fig. 7.1 Shaft work—work is produced with no volume change

where g is the acceleration due to gravity, z is the vertical height above some arbitrary horizontal level, and c is the velocity. Usually the initial and final values of z are equal, and the velocity is equal to zero, so eqn 7.3 reduces to $e = u$.

Turbines are essentially as shown in Fig. 7.1, though Fig. 7.2 shows a more realistic diagram. The gas flows over a series of fixed blades, also known as nozzles. These direct and accelerate the flow which then impinges on a set of moving blades. The blades are caused to move by the impact of the stream of gas against them. Similar machines also exist which work with a liquid. Figure 7.3 shows a diagram of one of the most simple to visualize: a Pelton wheel. The high-pressure liquid is forced through a nozzle, and a high speed jet is formed. This jet is directed against a set of 'buckets' mounted on a wheel. The impact of the liquid, and its change of momentum, causes a force on the buckets. This force causes the wheel to turn, and power can be generated.

7.3 The steady-flow energy equation

Consider the open system shown in Fig. 7.4. Fluid is flowing through it at a steady mass flow rate $\dot{m}$. The fluid enters with velocity c_1, at a pressure p_1 and specific volume v_1. The entry to the system is a height z_1 above a horizontal datum level. The corresponding quantities at exit from the system are: c_2, p_2, v_2, and z_2. The flow of heat to the system is $\dot{Q}$ and the rate the system does work on the surroundings is $\dot{W}_s$. The heat flow and the work flow (the subscript s indicates it is shaft work) are actually powers, and have the units joules/second or watts.

The closed-system version of the first law $Q - W = \Delta u$ cannot be used for the steady-flow system in Fig. 7.4. However, a deformable boundary can be designed which encloses a closed system, therefore

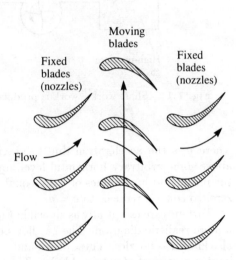

Fig. 7.2 Turbine showing rows of fixed blades (nozzles) and moving blades

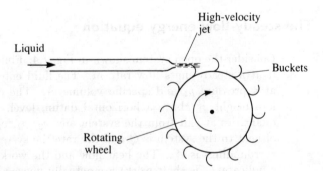

Fig. 7.3 The Pelton wheel: a turbine device driven by a liquid

no mass crosses the boundary. Figure 7.5 shows the initial and final system boundaries with a time interval of 1 second separating the initial and final states. The closed-system first law with the full definition of energy (eqn 7.2) can be applied to the fluid mass flowing in 1 second ($\dot{m}$). In this time the heat and work transfers are $\dot{Q}$ and $\dot{W}$. Hence:

$$\dot{Q} - \dot{W} = \dot{m}\Delta e \tag{7.4}$$

The work transfer from the system to the surroundings in 1 second consists of three parts:

1. The shaft work done by the system, $\dot{W}_{\mathrm{s}}$.

2. Work is needed to push the fluid into the system at A in Fig. 7.5. This work corresponds to a work done by the system of $-\dot{m}v_1p_1$. Here $-\dot{m}v_1$ corresponds to the decrease in volume of the system at A, and p_1 is the constant pressure at A.

3. Conversely work is done by the system on the surroundings at B of $+\dot{m}v_2p_2$.

Hence:

$$\dot{W} = \dot{W}_{\mathrm{s}} - \dot{m}v_1p_1 + \dot{m}v_2p_2 \tag{7.5}$$

Substituting eqn 7.5 into eqn 7.4 gives:

$$\dot{Q} - \dot{W}_{\mathrm{s}} + \dot{m}v_1p_1 - \dot{m}v_2p_2 = \dot{m}\Delta e \tag{7.6}$$

Then substituting for e from eqn 7.3 into eqn 7.6 and rearranging gives:

$$\dot{Q} - \dot{W}_{\mathrm{s}} = \dot{m}\left(u_2 + p_2v_2 + gz_2 + \frac{c_2^2}{2} - u_1 - p_1v_1 - gz_1 - \frac{c_1^2}{2}\right) \tag{7.7}$$

Equation 7.7 can be simplified first by writing it in the form:

$$\dot{Q} - \dot{W}_{\mathrm{s}} = \dot{m}\Delta\left(u + pv + gz + \frac{c^2}{2}\right) \tag{7.8}$$

and then by using the symbol h for the common combination of variables $u + pv$ to give the usual form of the first law for steady-flow system or, as it is often known, **the steady-flow energy equation**:

$$\dot{Q} - \dot{W}_{\mathrm{s}} = \dot{m}\Delta\left(h + gz + \frac{c^2}{2}\right) \tag{7.9}$$

The enthalpy (h for specific enthalpy (J/kg) or H for total enthalpy (J)) arises from the derivation of the steady-flow energy equation. As

$h = u + pv$, and u, p, and v are all properties of state, then h is also a property of state. As with internal energy this means that enthalpy can normally be expressed as a function of two other properties of state, thus:

$$h = f(T, v) = f(T, p) = f(p, v) \qquad (7.10)$$

It is interesting to consider the relative sizes of the terms on the right-hand side of eqn 7.9. As will be seen later, a change in enthalpy of 1000 J/kg is quite a small change, corresponding for air to a temperature change of approximately 1 K. For the remaining terms, on the right-hand side of the equation, this is equivalent to:

1. A change in vertical height of about 100 m.

2. A change in the velocity from zero to about 45 m/s.

Thus the potential energy term can usually be neglected, and the kinetic energy term can often be neglected, so that eqn 7.9 becomes:

$$\dot{Q} - \dot{W}_s = \dot{m}\Delta h \qquad (7.11)$$

or the differential form:

$$dQ - dW_s = dh \qquad (7.12)$$

The omission of the $\dot{m}$ from the right-hand side indicates that this is the equation per unit mass flowing. This form of the steady-flow energy equation is often quoted, but the significance of the extra terms in the full equation—eqn 7.9—must always be considered.

7.4 Enthalpy and specific heat

The last chapter considered the relationship between internal energy and the specific heat at constant volume. A corresponding result can be obtained with enthalpy:

$$c_p = \left(\frac{\partial h}{\partial T} \right)_p \qquad (7.13)$$

This is a perfectly general relationship between c_p and h.

Since the internal energy of an ideal gas depends only on temperature, and since for an ideal gas:

$$h = u + pv = u + RT \qquad (7.14)$$

the enthalpy is also dependent only on temperature. This has a number of consequences for an ideal gas.

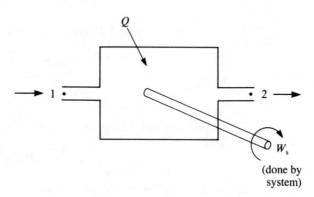

Fig. 7.4 A steady-flow system

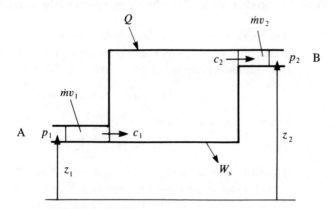

Fig. 7.5 Steady-flow system as a closed system: initial and final system boundaries

1. c_p can be written as:

$$c_p = \frac{dh}{dT} \tag{7.15}$$

2. h does not depend on pressure, so:

$$\left(\frac{\partial h}{\partial p}\right)_T = 0 \tag{7.16}$$

3. If c_p is a constant, eqn 7.15 can be integrated to give:

$$h_2 - h_1 = c_p(T_2 - T_1) \tag{7.17}$$

or simply:

$$h = c_p T \tag{7.18}$$

4. Differentiating eqn 7.14 gives:

$$\frac{dh}{dT} = \frac{du}{dT} + R \tag{7.19}$$

or substituting for dh/dT and du/dT:

$$c_p = c_v + R \tag{7.20}$$

So for air, $c_v = 718$ J/kg K and $R = 287$ J/kg K, so $c_p = 1005$ J/kg K.

7.5   Work transfer in a steady-flow reversible process

The work transfer in a reversible closed system process is given by $\int p\,dv$, but what is the corresponding equation for the shaft work in a steady-flow process? Rearranging eqn 7.5:

$$\dot{W}_s = \dot{W} + \dot{m}v_1p_1 - \dot{m}v_2p_2 \tag{7.21}$$

or for a unit mass of fluid:

$$W_s = W + v_1p_1 - v_2p_2 \tag{7.22}$$

The three terms on the right-hand side of eqn 7.22 can all be interpreted as areas on a p–v diagram (see Fig. 7.6). The resultant area is W_s, and therefore by this graphical argument, is seen to be given by:

$$W_s = -\int_1^2 v\,dp \tag{7.23}$$

An alternative, simpler derivation of eqn 7.23 will be given later.

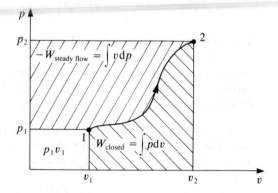

Fig. 7.6 Graphical interpretation of shaft work in a reversible steady-flow process

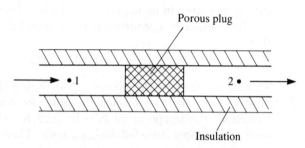

Fig. 7.7 The Joule–Thomson experiment

7.6 Examples: ideal gas systems

Example 7.1

Figure 7.7 shows the well-known Joule–Thomson[1] experiment. A gas flows along an thermally insulated tube and passes through a 'porous plug' in the tube. The porous plug could be a wad of cotton wool. Across the porous plug the pressure falls. If the gas is ideal, and the velocities low, what is the temperature change across the plug?

Solution

Since the velocities are low, and presumably the gravitational terms unimportant, the brief form of the steady-flow energy equation can be used. So for unit mass of the gas:

[1]William Thomson later became Lord Kelvin, so the experiment is sometimes known as the Joule–Kelvin experiment.

$$Q - W_\mathrm{s} = \Delta h \tag{7.24}$$

The tube is insulated and so $Q = 0$; no shaft work is done and so $W_\mathrm{s} = 0$; therefore there is no change in enthalpy. For an ideal gas the enthalpy depends only on temperature, and so for an ideal gas there is no change in temperature in this throttling process. This experiment is a sensitive test for an ideal gases.

Real gases do register some temperature change during this simple test. The size of the temperature change can be found from the Joule–Thomson coefficient, μ, which is defined by:

$$\mu = \left(\frac{\partial T}{\partial p} \right)_h \tag{7.25}$$

Equation 7.25 is used because the change of temperature with pressure is measured in an experiment which is designed to take place at constant enthalpy. For nitrogen at 1 bar and 25 °C it has been found that

$$\mu = +0.22 \text{ K/bar}$$

This means that as the pressure falls during the experiment (as it must to force the gas through the plug) for every 1 bar of pressure reduction, the temperature falls by 0.22 K. Almost every real gas shows this temperature fall during a Joule–Thomson experiment. The exceptions are hydrogen and helium which show temperature rises. Even these gases will show temperature falls if the gases are pre-cooled sufficiently. The temperature at which $\mu = 0$ is called the **inversion temperature**. The inversion temperature is actually a function of the pressure, but the maximum value of the inversion temperature is about 600 K for air, 200 K for hydrogen, and of order 50 K for helium. If a gas is cooled below its maximum inversion temperature, then it will show a temperature fall during a Joule–Thomson experiment.

Example 7.2   Air from a large reservoir at 298 K and 2 bar expands adiabatically through an insulated nozzle to the atmosphere at 1 bar (see Fig. 7.8). The nozzle exit temperature is 255 K. Calculate the air velocity at the nozzle exit.

Solution   It is clear that the full steady-flow energy equation including the velocity terms must be used. For unit mass of gas:

$$Q - W_\mathrm{s} = \Delta\left(h + gz + \frac{c^2}{2} \right) \tag{7.26}$$

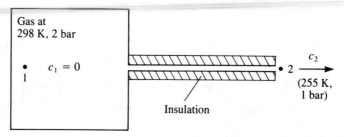

Fig. 7.8   Example 7.2: air flow in a nozzle

In this experiment, there is no heat transfer to or from the flow, there is no shaft work, and no change in vertical height, so eqn 7.26 becomes:

$$h_1 + \frac{c_1^2}{2} = h_2 + \frac{c_2^2}{2} \qquad (7.27)$$

where points 1 and 2 are shown in Fig. 7.8. The velocity at point 1 is zero, and h can be replaced by $c_p T$ giving:

$$c_2 = \sqrt{2c_p(T_1 - T_2)} \qquad (7.28)$$

Finally, substituting values into eqn 7.28 gives:

$$c_2 = \sqrt{2 \times 1005 \times (298 - 255)} = 294 \text{ m/s}$$

Later it will be shown that this nozzle flow is not a reversible process; however, the steady-flow energy equation is applicable to any steady-flow process: reversible or irreversible.

7.7   Examples: steam–water systems, the Rankine cycle

Example 7.3

A steady-flow steam–water cycle consists of four processes. The system states between these processes are defined in Table 7.1. The processes themselves are defined in Table 7.2. Figure 7.9 shows a block diagram of the whole system. The work and heat transfers in each process are to be calculated, as are the overall thermal efficiency of the cycle and the work ratio.

Two things should be noted from Tables 7.1 and 7.2:

1. The 'boiler' is, strictly, mis-named. The heating process in the boiler can be divided into three parts:

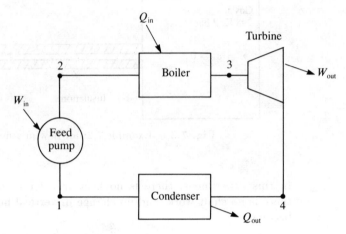

Fig. 7.9 Block diagram of the system for the steam–water example 7.3

Table 7.1 Definition of the states in the steam–water example 7.3

Point	Definition
1	saturated water at 20 °C (therefore $p_{sat} = 0.0233$ bar)
2	sub-cooled water at 100 bar and $\simeq 20°$C
3	superheated steam at 100 bar and 500°C
4	wet steam at 0.0233 bar and a dryness(x) of 0.8

Table 7.2 Definition of the processes in the steam–water example 7.3

Process	Definition
$1 \rightarrow 2$	adiabatic pump ($Q = 0$)
$2 \rightarrow 3$	'boiler' (no shaft work, $W_s = 0$)
$3 \rightarrow 4$	adiabatic turbine ($Q = 0$)
$4 \rightarrow 1$	condenser (no shaft work, $W_s = 0$)

Table 7.3 Values for v and h for saturated water and steam, to 221 bar

p (bar)	t_s (°C)	v_f (dm³/kg)	v_g (dm³/kg)	h_f (kJ/kg)	h_{fg} (kJ/kg)	h_g (kJ/kg)
0.00611	0.01	1.0002	206163	0.0	2501.6	2501.6
0.0233	20.00	1.0017	57838	83.9	2454.3	2538.2
1.00	99.63	1.0434	1693.7	417.5	2257.9	2675.4
10.00	179.88	1.1274	194.30	762.6	2013.6	2776.2
100.00	310.96	1.4526	18.04	1408.1	1319.6	2727.7
221.20	374.15	3.1700	3.17	2107.4	0.0	2107.4

(a) the sub-cooled water is heated up to the saturation temperature, here 311°C;

(b) the saturated water is evaporated at constant temperature, to give saturated steam;

(c) the saturated steam is superheated to the final temperature of 500°C.

2. Each of the four processes has either $Q = 0$ or $W_s = 0$. The heat or work transfers for each process per unit mass can be calculated from $Q - W_s = \Delta h$, so here:

$$Q \text{ or } -W_s = \Delta h \tag{7.29}$$

Hence all the heat and work transfers can be calculated from a knowledge of the enthalpies at each of the points 1, 2, 3, and 4 in Fig. 7.9. The assumption has been made here that the potential energy and the kinetic energy terms can be omitted from the steady-flow energy equation.

Solution Steam tables contain information about enthalpy presented in a similar way to the information about internal energy. Table 7.3 shows an extract from the saturation table. Again in these tables h_f is the saturation water enthalpy in kJ/kg, and h_g is the saturation steam enthalpy in kJ/kg. The quantity h_{fg} is simply:

$$h_{fg} = h_g - h_f \tag{7.30}$$

and is presented only because it is arithmetically useful when calculating the enthalpy of wet steam (see below). Two of the required enthalpies come directly from the steam tables (points 1 and 3), one requires a trivial calculation (point 4), and the remaining one (point 2) requires some thought.

Table 7.4   Values for v, u, and h for superheated steam, to 220 bar and 500 °C

p (bar) $(t_s\ (°C))$			Temperature (°C)				
		t_s	100	200	300	400	500
1.00	v (dm³/kg)	1693.7	1695.5	2172.3	2638.7	3102.5	3565.3
(99.60)	u (kJ/kg)	2506.0	2506.7	2658.2	2810.6	2968.0	3131.6
	h (kJ/kg)	2675.4	2676.2	2875.4	3074.5	3278.2	3488.1
10	v (dm³/kg)	194.30		205.92	257.98	306.49	353.96
(179.9)	u (kJ/kg)	2581.9		2620.9	2794.1	2957.9	3124.3
	h (kJ/kg)	2776.2		2826.8	3052.1	3264.4	3478.3
100	v (dm³/kg)	18.041				26.41	32.76
(311.0)	u (kJ/kg)	2547.3				2835.8	3047.0
	h (kJ/kg)	2727.7				3099.9	3374.6

1. Point 1 is saturated water at 20°C (therefore $p_{sat} = 0.0233$ bar). From the saturation Table 7.3 the result is:

$$h_1 = 83.9 \text{ kJ/kg} \tag{7.31}$$

2. Point 2 is sub-cooled water at 100 bar and $\simeq 20°C$. Tables for sub-cooled water (water below its saturation temperature) are not commonly available, but the enthalpy can be calculated. The reversible steady-flow work compressing the water from state 1 to state 2 can be calculated from:

$$W_s = - \int_1^2 v\,dp \tag{7.32}$$

The apparent problem now is that the exact relationship between specific volume and pressure during the compression process in the pump is not known. However to a good approximation, liquid water is incompressible and so v is constant and equal to approximately 0.001 m³/kg in this case. So eqn 7.32 becomes:

$$\begin{aligned} W_s &= -v(p_2 - p_1) = -0.001 \times (10^7 - 2330) \\ &= -9997 \text{ J/kg} \simeq -10.0 \text{ kJ/kg} \end{aligned} \tag{7.33}$$

Note that in eqn 7.33 the pressures were converted from bar to Pa, and the result for work has units J/kg. The actual pump is adiabatic and so:

$$h_1 - h_2 = -10.0 \text{ kJ/kg} \tag{7.34}$$

or

$$h_2 = 93.9 \text{ kJ/kg} \tag{7.35}$$

This derivation for h_2 depends on the pump being approximately reversible. Note that it also implies that although enthalpy for sub-cooled water, like the specific volume, can be expressed as $f(T, p)$, the enthalpy is a strong function of temperature, but only a very weak function of pressure. Increasing the pressure by a factor of 4000 only changes the enthalpy slightly.

3. Point 3 is superheated steam at 100 bar and 500°C. From Table 7.4 the enthalpy is:

$$h_3 = 3374.6 \text{ kJ/kg} \tag{7.36}$$

4. Point 4 is wet steam at 0.0233 bar and dryness (x) 0.8. In general the enthalpy of wet steam is calculated from:

$$h = (1 - x)h_f + xh_g \tag{7.37}$$

Equation 7.37 can be rearranged into a form which makes the calculations slightly quicker:

$$h = h_f + x(h_g - h_f) = h_f + xh_{fg} \tag{7.38}$$

Looking up the values of h_f and h_{fg} from the saturation Table 7.3, eqn 7.38 for point 4 becomes:

$$h_4 = h_f + xh_{fg} = 83.9 + 0.8 \times 2454.3 = 2047.3 \text{ kJ/kg} \tag{7.39}$$

The enthalpies at all the points in the cycle are now known, and eqn 7.29 can be used to calculate the separate heat and work transfers for the individual processes:

1. Pump $(1 \rightarrow 2)$:

$$W_s = h_1 - h_2 = 83.9 - 93.9 = -10.0 \text{ kJ/kg}$$

2. Boiler $(2 \rightarrow 3)$:

$$Q = h_3 - h_2 = 3374.6 - 93.9 = 3280.7 \text{ kJ/kg}$$

3. Turbine $(3 \rightarrow 4)$:

$$W_s = h_3 - h_4 = 3374.6 - 2047.3 = 1327.3 \text{ kJ/kg}$$

4. Condenser ($4 \rightarrow 1$):

$$Q = h_1 - h_4 = 83.9 - 2047.3 = -1963.4 \text{ kJ/kg}$$

The signs indicate that work is put into the cycle in the pump, and work comes out of the cycle in the turbine. Heat is put into the cycle in the boiler, and heat comes out of the cycle in the condenser. From these quantities the net work out and the net heat in can be calculated:

$$\text{net work out} = 1327.3 - 10.0 = 1317.3 \text{ kJ/kg}$$

$$\text{net heat in} = 3280.7 - 1963.4 = 1317.3 \text{ kJ/kg}$$

These are necessarily equal as $(\Delta h)_{\text{cycle}} = 0$. The thermal efficiency, η_{th} and the work ratio, r_{w} can also now be calculated:

$$\eta_{\text{th}} = \frac{\text{net work out}}{\text{heat in}} = \frac{1317.3}{3280.7} = 0.401$$

$$r_{\text{w}} = \frac{\text{net work out}}{\text{work out}} = \frac{1317.3}{1327.3} = 0.992$$

As stated in Chapter 5, both these quantities should be as near 1 as possible. The work ratio is, by any standards, nearly equal to one, but the thermal efficiency is not. The actual significance of an efficiency of around 40% will be discussed later in the context of the second law of thermodynamics. It is, however, quite a reasonable value for the thermal efficiency. Later it will also be shown that the turbine in this question is irreversible, but this only contributes in a small way to the apparently low efficiency.

This cycle is called the **Rankine cycle**. The cycle can be plotted on a p–v diagram: Fig. 7.10 shows a sketch of such a diagram. The Rankine cycle is one by which heat is taken from burning fuel (oil, gas, or coal) and partially converted into work which is then used to generate electricity. Even the most efficient electricity generating stations using this cycle do not have overall efficiencies much above 40%. The heat which is not converted into work is rejected from the cycle. In the example above the heat is rejected in the condenser at a temperature of only 20°C. Although there is a large quantity of heat, at this low temperature it is useless and valueless. If the exit pressure of the turbine and the condenser pressure were raised to 1 bar, then the heat would be rejected at a more useful temperature—in this case 100°C, but the work output would be reduced. However, this heat

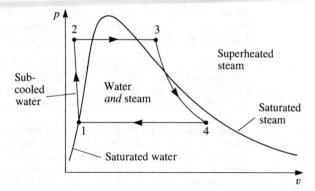

Fig. 7.10   The Rankine cycle sketched on a p–v diagram for steam–water

could be used for heating buildings in **district heating** schemes. The problem then is the organisation of the distribution of the large quantities of heat and the large capital investment such schemes require. Cycles which produce both power and useful heat in this way are known as **combined heat and power (CHP)** cycles.

7.8   Problems

7.1 Air from a large reservoir at 300 K and 1.5 bar expands adiabatically through a nozzle to 1 bar. The air velocity at the nozzle exit is 250 m/s. Calculate the air temperature.

7.2 An adiabatic air turbine operates with an air supply at 10 bar and 400 K. The inlet air velocity is 30 m/s and the outlet velocity is 200 m/s. The outlet conditions are 1 bar and 250 K. Calculate the work produced per kg air flowing.

7.3 A adiabatic throttling process takes fluid at specified inlet conditions, and reduces the pressure to 1 bar. For the following inlet conditions, calculate the state of the fluid after the throttling:

(a) saturated water at 50 bar;

(b) steam–water at 10 bar, and a dryness x of 0.99;

(c) air at 2 bar and 500°C.

7.4 Using the results of question 7.3, suggest a method for measuring the dryness of almost dry steam. Why cannot the method be used for very wet steam?

7.5 In a Rankine cycle, the steam at the boiler outlet is saturated at 10 bar. At the end of the turbine the conditions are 0.04 bar and a dryness of 0.80. At the end of the condenser the water is saturated. Calculate the thermal efficiency and the work ratio of the cycle.

7.6 The Rankine cycle in question 7.5 is now modified so that the turbine exit conditions are 1 bar and $x = 0.90$. Calculate the thermal efficiency. Explain why this cycle, although it has a lower efficiency than the cycle in question 7.5, might be more 'efficient' in some sense.

7.7 A compressor which supplies air at a pressure of 6 bar and 350 K is connected to an insulated, rigid tank of volume 3 m³ which initially contains air at 1 bar and 300 K. Calculate the mass of air which flows into the tank when its pressure reaches 6 bar, and calculate the final temperature of the air in the tank. [Hint: think carefully—is this a question about steady flow? If it is not, you should not be using the steady-flow energy equation.]

7.8 Repeat question 7.7 with steam. The compressor supplies saturated steam at 6 bar, and the tank initially contains saturated steam at 1 bar.

8

Heat engines, heat pumps, and the second law of thermodynamics

8.1 Key points of this chapter

- A generalized definition of thermal efficiency is given in terms of heat and work transfers, and in terms of heat transfer only. (Section 8.2)

- The concept of hot and cold reservoirs of heat is introduced for use in the analysis of heat engines. (Section 8.2)

- A reversible heat engine can be run 'backwards' and convert work into heat, the heat being liberated at a high temperture. This is the principle of the operation of a heat pump and a refrigerator. (Section 8.2)

- The index of performance for heat pumps and refrigerators, analogous to the thermal efficiency of a heat engine, is the coefficient of performance. (Section 8.3)

- Various statements, apparently very different, of the second law of thermodynamics are given. (Section 8.4)

- The equivalence of two of these statements, those due to Planck and Clausius, are demonstrated. (Section 8.5)

8.2 Heat engines

A heat engine is the generalized name for a machine which converts heat into work. It will be evident from the example of the Rankine cycle in Chapter 7, that heat and work are not just different manifestations of energy. There are very distinct differences. This is clearly seen in the definition of the thermal efficiency, η_{th} of a cycle:

$$\eta_{th} = \frac{\text{net work out}}{\text{heat in}} \tag{8.1}$$

The efficiency in eqn 8.1 is gauged in terms of the **net** work out. The work produced in the cycle and the work consumed in the cycle, or positive and negative work, have equal 'values'. The value of the work out can simply be obtained by the arithmetic sum of all the work transfers. However, heat is very different. All heat does not have the same value. Equation 8.1 contains only the heat taken into the cycle, and the heat given out of the cycle is ignored. This is because the heat given out is usually assumed to have no value. Another way to look at the thermal efficiency is that:

$$\eta_{th} = \frac{\text{what you want}}{\text{what you have to pay for}} \tag{8.2}$$

In eqn 8.2 'what you want' is the net work produced by the cycle, and 'what you have to pay for' is the heat put into the cycle—again counting the heat given out as without value. For this reason, a generalized diagram of a heat engine is often used, as shown in Fig. 8.1. The essential features of this generalized description are as follows.

1. The work input and the work output have been amalgamated to be shown as simply the net work output.

2. The heat input and the heat output are shown separately.

3. The heat input to the cycle is assumed to be supplied from a source of high temperature heat. This source of heat is termed a **heat reservoir**, and this is the **hot reservoir**.

4. The heat output from the cycle is assumed to go to a **cold reservoir**.

In terms of the symbols in Fig. 8.1, the thermal efficiency is given by:

$$\eta_{th} = \frac{W}{Q_1} \tag{8.3}$$

or, since it is evident from Fig. 8.1 that $W = Q_1 - Q_2$, eqn 8.3 can be rewritten as:

$$\eta_{th} = \frac{Q_1 - Q_2}{Q_1} = 1 - \frac{Q_2}{Q_1} \tag{8.4}$$

Figure 8.2 shows a version of Fig. 8.1 for the steam–water example from Chapter 7. Note that for every 100 units of heat from the hot

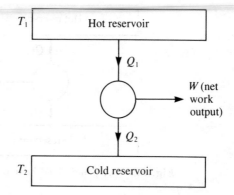

Fig. 8.1   A heat engine: generalized diagram

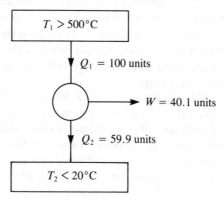

Fig. 8.2   A generalized heat engine: steam–water example from Chapter 7

reservoir (Q_1), 40.1 units of work are produced (W) giving an efficiency of 40.1%. Note also that the temperature of the hot reservoir (T_1) must be greater than or equal to 500°C and the temperature of the cold reservoir (T_2) must be less than or equal to 20°C.

If all the processes making up the cycle are reversible, and the heat is transferred in and out of the cycle reversibly, then the heat engine is a reversible heat engine. If then the reversible heat engine is run backwards so that work is put into the cycle and heat is rejected at the high temperature, that we have a **heat pump** or a **refrigerator**.

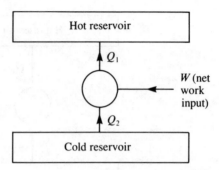

Fig. 8.3   A generalized heat pump or refrigerator

8.3   Heat pumps and refrigerators

Figure 8.3 shows a heat pump or a refrigerator. This diagram is very similar to Fig. 8.1. The difference is now that the heat is 'flowing uphill', that is from the cold reservoir to the hot reservoir. To accomplish this, it is necessary to add work to the cycle. So the direction of all the heat and work transfers have been reversed from the position with the heat engine.

If Fig. 8.3 represents a refrigerator, then the cold reservoir temperature (T_2) would be below room temperature, and the hot reservoir temperature (T_1) would be above room temperature. T_2 would be below room temperature to ensure that the inside of the refrigerator remained cool; T_1 would be above room temperature to ensure that the heat rejected by the cycle could be transferred back to the surroundings of the refrigerator. Figure 8.4 shows the heat flows for a refrigerator in more detail.

Obviously the thermal efficiency cannot be used to characterize a refrigerator, but eqn 8.2 can be used. In a refrigerator 'what you want' is heat removed from the cold reservoir, say the freezing compartment of the refrigerator, Q_2 in this case. 'What you have to pay for' is the work that has to be put into the cycle, here W. The ratio of these, as in eqn 8.2 is no longer an efficiency; it is known as the **coefficient of performance** or COP. So, for a refrigerator:

$$(\text{COP})_\text{R} = \frac{Q_2}{W} \tag{8.5}$$

In a heat pump the objective is rather different: it is to take otherwise useless heat—maybe from the water in a river—and raise the temperature of the heat to a useful level. If the outlet temperature

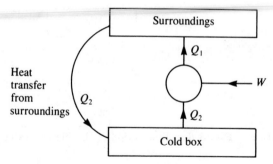

Fig. 8.4   The heat flows in a refrigerator

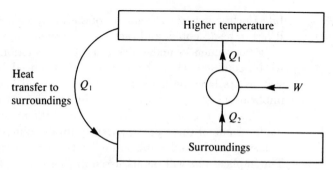

Fig. 8.5   The heat flows in a heat pump

is high enough the output heat can be used to heat a building. Figure 8.5 shows the heat flows in a heat pump installation. Since the objective is different, 'what you want' has also changed. It is now heat output that is required, that is Q_1. However, the net work input W remains the quantity that has to be paid for. Hence, the coefficient of performance for a heat pump is:

$$(\text{COP})_{\text{HP}} = \frac{Q_1}{W} \tag{8.6}$$

Since from Fig. 8.3 it is obvious that $Q_2 + W = Q_1$, there is obviously a simple relationship between these two versions of the coefficient of performance. In fact:

$$(\text{COP})_{\text{HP}} = (\text{COP})_{\text{R}} + 1 \tag{8.7}$$

The concept of a heat engine and a heat pump is a useful one in the consideration of the second law of thermodynamics.

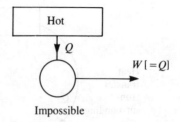

Fig. 8.6 Illustration of the Planck statement of the second law

8.4 The second law of thermodynamics

Like the first law, the second law of thermodynamics cannot be proved. It has to be accepted, and then its consequences tested. Only because these consequences make sense, and the predictions are followed, can we be confident that the second law is indeed true. Many grand statements have been made about the second law; for example the implications of the law to philosophy and the nature of time.

What is the second law? Various statements of it have been made—some of these are apparently totally different, and from none of them are the full implications of the second law at all apparent. The implications will be discussed in Chapter 9. However the following are three statements of the second law.

A In the neighbourhood of each equilibrium state of a thermodynamic system there exist states that are inaccessible by adiabatic processes. This is the statement of the second law by Carathéodory (1909).

B It is impossible to construct a system operating in a cycle which extracts heat from a reservoir and does an equivalent amount of work on the surroundings. This is the statement of the second law by Planck (1927), and illustrated in Fig. 8.6.

C It is impossible to construct a system operating in a cycle which transfers heat from a cooler body to a hotter body without work being done on the system by the surroundings. This is the statement of the second law by Clausius (c.1850), and illustrated in Fig. 8.7.

Statement A is mathematical and abstract and will not be considered here. Notice though how different statements of the second law can actually be. Here, attention will be focussed on statements B and

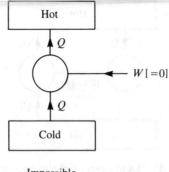

Impossible

Fig. 8.7   Illustration of the Clausius statement of the second law

C, due to Planck and Clausius. These are written in the language of heat engines: they are statements about systems, cycles, heat, and work. They appear very limited statements without, at first sight, saying anything very profound. The far-reaching consequences of these statements will be considered in Chapter 9. First however it is necessary to demonstrate the equivalence of statements B and C, that is to demonstrate that they are consistent. They are either both true or both false.

8.5   Equivalence of the Planck and Clausius formulations of the second law

The equivalence will be demonstrated by assuming that if one statement is false, then the other is false also. This is done twice. First statement B is assumed to be false, and then statement C is assumed to be false.

1. Suppose B, the Planck statement, is not true. If B is not true, then heat can entirely be converted into work, as shown in the left-hand part of Fig. 8.8. The work produced is used in a Clausius-type process where heat is transferred from a colder to a hotter reservoir. However, the net result of these two devices (see the right hand side of Fig. 8.8) is itself a process that violates statement C—the Clausius version of the second law.

   So if B is not true, then C is also not true.

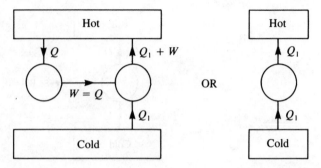

Fig. 8.8 Consequence of the assumption that the Planck statement (B) is not true

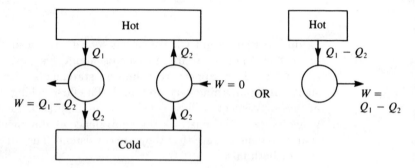

Fig. 8.9 Consequence of the assumption that the Clausius statement (C) is not true

2. Suppose C, the Clausius statement, is not true. If C is not true, then heat can entirely be transferred from a colder reservoir to a hotter reservoir without any work being required as shown in the left hand part of Fig. 8.9. The heat supplied by the colder reservoir is alternatively supplied by a Planck-type process where heat flows from a hotter to a colder reservoir, and some work is produced. However, the net result of these two devices (see the right hand side of Fig. 8.9) is itself a process that violates statement B—the Planck version of the second law.

So if C is not true, then B is also not true.

Hence either B and C are both true, or B and C are both false. We have not proved that either statement is actually true, only that the statements are consistent. Chapter 9 looks at the consequences of assuming that the second law in these forms is true. We will look at four main corollaries of the second law, and also find the most efficient way of converting heat into work—the Carnot cycle.

9

Corollaries of the second law

9.1 Key points of this chapter

This chapter is concerned with the corollaries or consequences of the second law. Four main consequences are discussed here together with one of the most important thermodynamic cycles: the Carnot cycle.

- The first corollary is that the reversible heat engine is the most efficient heat engine operating between two temperatures. (Section 9.2)

- The second corollary is that there is a purely thermodynamic definition of temperature based on the concept of a reversible heat engine. (Section 9.3)

- The definition of thermodynamic temperature adopted is that in a reversible heat engine the heat flows are proportional to the absolute temperature of the heat reservoirs. (Section 9.3)

- The Carnot cycle is the most efficient heat engine operating between two temperature reservoirs. (Section 9.4)

- It is demonstrated that the thermodynamic temperature and the ideal gas temperature are identical. (Section 9.4)

- The Clausius inequality puts a necessary restriction on the integrated heat transfer round a cycle. (Section 9.5)

- The Clausius inequality leads to the existence of the thermodynamic property entropy. (Section 9.6)

- Entropy is a true property of state, depending only on the state of the system and not on the particular path of any process. (Section 9.6)

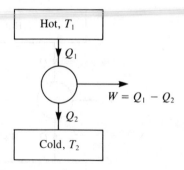

Fig. 9.1   Generalized heat engine

9.2   The reversible heat engine is the most efficient heat engine

The first corollary of the second law considered here concerns a reversible heat engine, and the fact that for a cycle operating between a given hot reservoir and a given cold reservoir, the reversible heat engine is the most efficient heat engine. First, a reminder of the basic heat engine is given in Fig. 9.1 and the thermal efficiency of the engine η_{th} is defined by:

$$\eta_{th} = \frac{W}{Q_1} \qquad (9.1)$$

The reversible heat engine by definition can be reversed. If this is done the energy flows in Fig. 9.1 are unchanged in magnitude but reversed in direction. The supposition that the reversible heat engine is the most efficient one is proved by assuming it is not true and that an engine X exists which is more efficient. Figure 9.2 shows two heat engines coupled together. Engine X is operating as a heat engine, that is taking heat from the hot reservoir, converting part of it into work, and rejecting some heat to the cold reservoir. Engine R is the reversible heat engine, running reversed as a heat pump, that is taking in heat from the cold reservoir, and taking in work to reject a relatively large amount of heat to the hot reservoir.

If, as shown in Fig. 9.2, both X and R take and reject the same amount of heat Q_1 at the hot reservoir, then engine X will give out more work W_X than is required by R, W_R. This is because engine X is assumed to be more efficient, that is $\eta_X > \eta_R$. Because the same amount of heat is given up to the hot reservoir as is taken from it, that is Q_1, the hot reservoir can be removed (see Fig. 9.3). The situation now is that some heat is being removed from the cold reservoir and is entirely being converted into work. This violates the

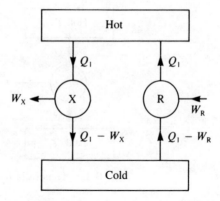

Fig. 9.2 Heat engine X and the reversed reversible heat engine R

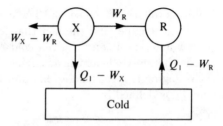

Fig. 9.3 Net result of the combination X and R reversed

Planck statement of the second law: heat cannot entirely be turned into work. The original assumption that X is more efficient than R must therefore be false, and the reversible heat engine must be the most efficient heat engine.

9.3 The thermodynamic temperature

The second corollary of the second law considered here concerns the thermodynamic temperature. The corollary is that a temperature, the 'thermodynamic temperature', can be defined without reference to the properties of any substance, and that there is an absolute zero to this temperature scale. The efficiency of the reversible heat engine in Fig. 9.4 can be expressed as:

$$\eta_{th} = 1 - \frac{Q_2}{Q_1} \tag{9.2}$$

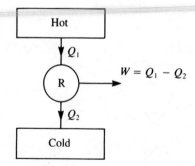

Fig. 9.4 Reversible heat engine

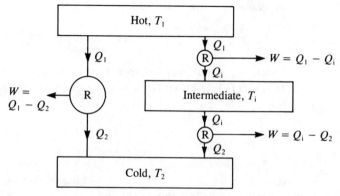

Fig. 9.5 Three reversible heat engine operating between hot, cold, and intermediate temperature reservoirs

Now the efficiency does not depend on the actual magnitude of Q_1 and Q_2 but only on their ratio. The efficiency must therefore be dependent on some property of reservoirs, and that property is the temperature; however, the temperature is actually to be defined. Hence we can write that the efficiency depends on the temperatures of the two reservoirs in some way:

$$\eta_{th} = 1 - f(T_2, T_1) \tag{9.3}$$

Can we say anything about the form of this function $f(T_1, T_2)$? Consider the situation with three reversible heat engines shown in Fig. 9.5. A reservoir at an intermediate temperature, here called T_i has been placed between the hot and the cold reservoirs. Since this function

$f(T_1, T_2)$ is always the same function, the following equations can be written:

$$\frac{Q_2}{Q_1} = f(T_2, T_1) \qquad (9.4)$$

$$\frac{Q_i}{Q_1} = f(T_i, T_1) \qquad (9.5)$$

and

$$\frac{Q_2}{Q_i} = f(T_2, T_i) \qquad (9.6)$$

Now since:

$$\frac{Q_2}{Q_1} = \frac{Q_i}{Q_1} \times \frac{Q_2}{Q_i} \qquad (9.7)$$

we can also write that:

$$f(T_2, T_1) = f(T_i, T_1) \times f(T_2, T_i) \qquad (9.8)$$

From eqn 9.8 it can be seen that $f(T_2, T_1)$ must be of the form $f(T_2)/f(T_1)$, but no further help about the actual form of the function is available. The simplest function satisfying this criterion is that:

$$\frac{Q_2}{Q_1} = \frac{T_2}{T_1} \qquad (9.9)$$

though other choices are also possible. In fact eqn 9.9 is the choice actually made and this is the definition of the thermodynamic temperature. Therefore, in a reversible heat engine the ratio of the heat flows to and from the two heat reservoirs is equal to the ratio of the absolute temperatures of the reservoirs. In the next section it will be shown that this thermodynamic temperature is identical to the ideal gas temperature. One immediate consequence of the choice of function relating heat flow to temperature made in eqn 9.9 is that the efficiency of a reversible heat engine is given by:

$$\eta_{th} = 1 - \frac{T_2}{T_1} \qquad (9.10)$$

Another immediate consequence is that the Planck statement of the second law says that $Q_2 > 0$, and so therefore we must also have that $T_2 > 0$. Therefore the temperature, measured on this thermodynamic scale, must always be greater than zero. This **absolute zero of temperature** is, moreover, unattainable since we have always the requirement that $Q_2 > 0$.

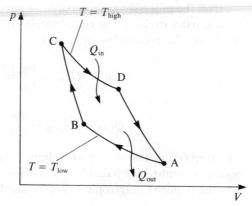

Fig. 9.6 The Carnot cycle

9.4 The Carnot cycle

Equation 9.10 defines the maximum efficiency of a heat engine operating between the two temperatures T_1 and T_2. But how can this efficiency, even in theory, be obtained? The important considerations are as follows.

1. All the processes inside the cycle must themselves be reversible.

2. All the heat transferred in and out of the cycle must be transferred reversibly. If the heat transfer is to be reversible, then the heat must be transferred to the cycle **at** the top temperature T_1, and the heat rejected must be transferred **at** the bottom temperature T_2.

An ideal gas cycle already considered has the property that all the processes in the cycle can in theory be reversible processes, and that all the heat transfer takes place at the highest and lowest temperatures of the cycle (see Fig. 9.6). This is the Carnot cycle: two reversible isothermal processes and two reversible adiabatics. The processes B → C and D → A are adiabatic processes and therefore there is no heat transfer along these parts of the cycle.

The processes A → B and C → D are isothermals, and so from Chapter 6, it will be recalled that there is heat transfer during these processes. The process A → B takes place at the lowest temperature in the cycle, say T_{low}, and the process C → D takes place at the highest temperature, say T_{high}.

If all the processes are reversible we have all the correct ingredients for a **Carnot cycle**. From the equations developed in Chapter 6, the two heat transfers can be evaluated:

$$Q_{AB} = RT_{low} \ln \frac{V_B}{V_A} = -Q_{out} \tag{9.11}$$

$$Q_{CD} = RT_{high} \ln \frac{V_D}{V_C} = Q_{in} \tag{9.12}$$

The connection between these volume ratios can be found from the equations connecting p and V at subsequent points in the cycle. Here n is the required polytropic index for adiabatic operation:

$$p_A V_A = p_B V_B \tag{9.13}$$

$$p_B V_B^n = p_C V_C^n \tag{9.14}$$

$$p_C V_C = p_D V_D \tag{9.15}$$

$$p_D V_D^n = p_A V_A^n \tag{9.16}$$

Eliminating the four pressures from the four eqns 9.13 to 9.16 then gives:

$$\frac{V_D}{V_C} = \frac{V_A}{V_B} \tag{9.17}$$

Finally substituting eqn 9.17 into eqn 9.11 and 9.12 gives:

$$\frac{Q_{in}}{Q_{out}} = \frac{T_{high}}{T_{low}} \tag{9.18}$$

This equation is identical to eqn 9.9. Equation 9.18 was derived from ideal gas equations, and eqn 9.9 was derived from the definition of the thermodynamic temperature. The important conclusion is that the thermodynamic temperature is the same as the ideal gas temperature.

This equality of the temperature scales is a consequence of the decision to relate heat flows to temperature by eqn 9.9. Alternative choices would not have made the temperature scales the same. In fact Kelvin considered the use of an alternative form of eqn 9.9:

$$\ln \frac{Q_2}{Q_1} \propto \theta_2 - \theta_1 \tag{9.19}$$

On this temperature scale, absolute 'zero' would be $\theta = -\infty$.

The efficiency of 40.1% obtained in the Rankine cycle in Chapter 7 can now be examined. In this cycle:

$$T_{\text{high}} = 500°C = 773\text{ K} \quad \text{and} \quad T_{\text{low}} = 20°C = 293\text{ K}$$

and so for these temperatures, from eqn 9.10:

$$\eta_{\text{Carnot}} = 1 - \frac{293}{773} = 0.621$$

So the efficiency actually obtained is lower than the Carnot efficiency, but not all that much lower. The reasons why the efficiency is lower than the Carnot efficiency are as follows.

1. A minor reason is that the turbine in the Rankine cycle is not reversible, and so the maximum work is not obtained.

2. The major reason is that whilst the heat is really all rejected in the condenser at 20°C, the heat is added in the boiler at temperatures ranging from 20°C to 500°C.

However, the Carnot efficiency does provide a quick check on a calculation and can rule out impossible answers: the efficiency, in this example, cannot exceed 0.621.

9.5 The Clausius inequality

The third main corollary of the second law is the Clausius inequality, which is that when a closed system operates in a cycle:

$$\oint \frac{dQ}{T} \leq 0 \tag{9.20}$$

In fact if the cycle is reversible, the 'equals' sign applies: if it is irreversible, the 'less than' sign applies. To prove this, consider a system A, see Fig. 9.7, undergoing a cyclic process. During part of this cycle, the system A receives heat dQ_A in a part of the system where the temperature is T. The system does work dW_A on the surroundings. Imagine that the heat is supplied from an external reservoir of heat at a temperature T_0 via a reversible heat engine R which does

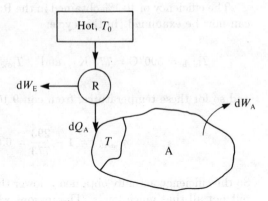

Fig. 9.7   System for proof of the Clausius inequality

work dW_E on the surroundings. From the thermodynamic definition of temperature:

$$\frac{dW_E + dQ_A}{dQ_A} = \frac{T_0}{T} \qquad (9.21)$$

or

$$dW_E = \frac{T_0}{T} dQ_A - dQ_A \qquad (9.22)$$

Integrating eqn 9.22 over a whole cycle of the process,

$$\oint dW_E = T_0 \oint \frac{dQ_A}{T} - \oint dQ_A \qquad (9.23)$$

Applying the first law to the system A gives:

$$\oint dQ_A = \oint dW_A \qquad (9.24)$$

and so eqn 9.23 becomes:

$$\oint dW_E = T_0 \oint \frac{dQ_A}{T} - \oint dW_A \qquad (9.25)$$

This equation is for the system A and the reversible heat engine R. This combined system gets heat only from the reservoir at T_0 and rejects no heat. Thus the work it does, from the second law, on its surroundings must be negative, that is:

$$\oint (dW_E + dW_A) \leq 0 \tag{9.26}$$

Substituting eqn 9.26 into eqn 9.25 gives:

$$\oint \frac{dQ_A}{T} \leq 0 \tag{9.27}$$

This is the Clausius inequality applied to the system A.

To see the significance of the equals sign or the less than sign in eqn 9.27, suppose that system A is undergoing a reversible cycle. R is already reversible. Hence the inequality is the same for the system operating forwards or backwards, but the sign of Q has changed. This can only be true if, for a reversible cycle:

$$\oint \frac{dQ_A}{T} = 0 \tag{9.28}$$

and then, the result:

$$\oint \frac{dQ_A}{T} < 0 \tag{9.29}$$

must be true for an irreversible cycle.

9.6 The thermodynamic property entropy

The fourth corollary of the second law is there exists a property of state, the entropy S, which is defined by:

$$dS = \frac{dQ_{rev}}{T} \tag{9.30}$$

where dQ_{rev} is the reversible heat transfer to the system (J), and T is the absolute temperature (K). The units of entropy S are J/K. The corresponding entropy per unit mass, the specific entropy s, has units J/kg K. The significance of entropy being a property of state is that the entropy change between two states only depends on the initial and final states and not on the exact path of the process. Hence the entropy change:

$$s_2 - s_1 = \int_1^2 \left(\frac{dQ}{T} \right)_{rev} \tag{9.31}$$

where 1 and 2 represent the initial and final states, does not depend on the path, only on the end states 1 and 2. The existence of this

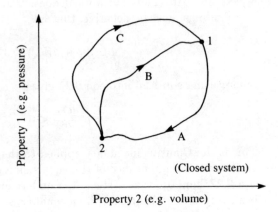

Fig. 9.8   Reversible processes A, B, and C connecting two states 1 and 2

property of state entropy can be proved by considering Fig. 9.8. Consider a cycle from point 1 to point 2 and back again. There are many such reversible cycles: one is made up from the reversible processes A and B, another is made up from the reversible processes A and C. For these cycles:

$$\int_A \frac{dQ}{T} + \int_B \frac{dQ}{T} = \int_A \frac{dQ}{T} + \int_C \frac{dQ}{T} = 0 \qquad (9.32)$$

since $\oint dQ/T = 0$ from the Clausius inequality for a reversible cycle. From eqn 9.32 by subtraction:

$$\int_B \frac{dQ}{T} = \int_C \frac{dQ}{T} \qquad (9.33)$$

and so $\int_1^2 dQ/T$ is independent of path, and can be written as $s_2 - s_1$ where s is a property of state and therefore in general a function of two other properties of state such as pressure and volume.

We have shown that for a reversible process $ds = \int dQ/T$ but what about an irreversible process? Consider a similar situation to that shown in Fig. 9.8 (see Fig. 9.9). A reversible cycle can still be made up from the reversible processes A and B. If we now assume that process C is irreversible, then an irreversible cycle can be made up from processes A and C. For the reversible cycle:

$$\int_A \frac{dQ}{T} + \int_B \frac{dQ}{T} = 0 \qquad (9.34)$$

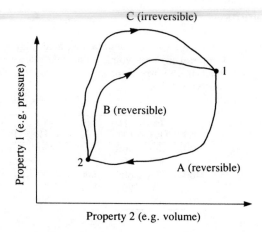

Fig. 9.9 Three processes A, B, and C connecting two states 1 and 2: A and B are reversible, C is irreversible

For the irreversible cycle, from the Clausius inequality $\oint dQ/T < 0$, so:

$$\int_A \frac{dQ}{T} + \int_C \frac{dQ}{T} < 0 \qquad (9.35)$$

From eqns 9.34 and 9.35 by subtraction replacing the labels B and C by 'rev' (for reversible) and 'irrev' (for irreversible) respectively:

$$\int_{rev} \frac{dQ}{T} > \int_{irrev} \frac{dQ}{T} \qquad (9.36)$$

or, in differential form:

$$ds > \left(\frac{dQ}{T}\right)_{irrev} \qquad (9.37)$$

The general result is that:

$$ds \geq \frac{dQ}{T} \qquad (9.38)$$

In eqn 9.38 the 'equals' sign applies to a reversible process and the 'greater than' sign applies to an irreversible process.

Chapter 10 considers further aspects of entropy: its use as a criterion of reversibility, what it represents on a molecular level, and the calculation of entropy in some simple examples.

9.7   Problems

9.1 Figure 9.1 shows a generalized heat engine operating between infinite temperature reservoirs. The heat flows and the work are positive when in the directions shown in Fig. 9.1. Complete the blanks in the following table.

T_1 (K)	T_2 (K)	Q_1 (J)	Q_2 (J)	W (J)	Possible or impossible Reversible or irreversible
400	200	500	?	250	?
500	200	800	?	550	?
400	?	800	500	?	irreversible
10	50	200	2000	?	?
200	300	?	?	-1000	reversible

9.2 A block of metal (mass m, specific heat C) is originally at a temperature T_1. A reversible heat engine is run between the block and the surroundings at T_0. Prove that the work extracted is:

$$W = mcT_0 \left(\frac{T_1}{T_0} - 1 - \ln \frac{T_1}{T_0} \right)$$

whether $T_1 > T_0$ or $T_1 < T_0$. Show that in either case the work is positive.

9.3 A block of metal (mass m, specific heat C) below ambient temperature is cooled from T_1 to T_2 by means of a reversible heat pump rejecting heat to the surroundings at T_0. Prove that the work necessary is

$$W = -mCT_0 \ln \frac{T_2}{T_1} + mC(T_2 - T_1)$$

If $m = 1$ kg, $C = 1$ kJ/kg K, and $T_0 = 300$ K find the work input necessary to cool the block by 1 K if the initial temperature T_1 is

(a) 299 K

(b) 2 K

9.4 Equation 9.19 can be used to define a temperature scale, $^\circ$T (T for Thomson) and the melting point of ice at 1 bar is defined as 0°T, and the boiling point of water at 1 bar as 100°T. Calculate the temperatures corresponding to 5000 K and -1000°T.

9.5 Explain why the following equations cannot be used in place of eqn 9.19:

$$\ln \frac{Q_2}{Q_1} = \frac{\theta_2}{\theta_1} \quad \text{and} \quad \ln \frac{Q_2}{Q_1} = \frac{\theta_2}{\theta_1} - 1$$

10

Entropy

10.1 Key points of this chapter

A process occurring at constant entropy is known as an **isentropic process**. Since entropy change is defined as $\int (dQ)_{\text{rev}}/T$, an isentropic change is both reversible and adiabatic.

- The entropy change of the universe (the system and the surroundings) is the criterion of reversibility. (Section 10.2)

- Entropy always increases (for an irreversible process) or remains constant (for a reversible process), never decreases. (Section 10.2)

- An entropy change for a process can be calculated for a reversible change. This is then the entropy change for any process between these initial and final states. (Section 10.3)

- Heat transfer across finite temperature differences is a cause of irreversibility. (Section 10.3)

- A number of equations representing the combined first and second laws of thermodynamics are very useful because they are valid for any substance and for any type of change. (Section 10.4)

- Using these equations for the combined laws, a convenient derivation of the shaft work in a reversible, steady-flow process is available. (Section 10.4)

- Equations for the entropy change in an ideal gas are derived. From these, the equation of an isentropic change can be derived. (Section 10.5)

- A sudden, unresisted expansion of a gas is an irreversible process. (Section 10.5)

• Calculations for steam–water systems are, in principle, no more difficult than for ideal gases, but the entropy has to be obtained from tabulated data. (Section 10.6)

10.2 Entropy as a criterion of reversibility

At the end of Chapter 9 it was concluded that:

$$ds \geq \frac{dQ}{T} \tag{10.1}$$

where the 'equals' sign referred to a reversible process and the 'greater than' sign referred to an irreversible process. This entropy change, ds was the entropy change of just the system. If now the surroundings are considered also, then this heat dQ is exchanged with the surroundings. If the heat is transferred **to** the system, then it is obviously transferred **from** the surroundings. If the heat is exchanged reversibly with the surroundings, then:

$$(ds)_{\text{surroundings}} = -\frac{dQ}{T} \tag{10.2}$$

If the process is reversible, then the entropy change of the system is dQ/T, and the total entropy change, the entropy change of the universe is zero:

$$(ds)_{\text{universe}} = (ds)_{\text{surroundings}} + (ds)_{\text{system}} = 0 \tag{10.3}$$

If the process is irreversible, then the entropy change of the system is greater than dQ/T, and the total entropy change of the universe is greater than zero:

$$(ds)_{\text{universe}} = (ds)_{\text{surroundings}} + (ds)_{\text{system}} > 0 \tag{10.4}$$

Hence, combining eqns 10.3 and 10.4:

$$(ds)_{\text{universe}} \geq 0 \tag{10.5}$$

where the 'equals' sign is the relevant one for a reversible process, and the 'greater than' sign is the relevant one for an irreversible process. Note that the entropy change of the universe cannot be less than zero. Thus entropy of the universe is always increasing. The statement that the entropy of the universe is **always** increasing can be made with confidence because all real processes are to some extent irreversible. Reversibility can only ever be approximately obtained. The fact that the entropy of the universe is always increasing has led to entropy being described as 'the arrow of time'. A film or video of any process can be shown backwards—starting at the present and ending with an event in the past. If everything about the process is known, the entropy can be calculated at any moment. A backwards running film could be detected because the entropy would appear to be decreasing. No real process is truly reversible because the entropy always increases.

A common mistake made is to remember that entropy is the criterion of reversibility and that entropy always tends to increase, but to forget that these statements apply only to the entropy of the system **and** the surroundings. It is quite feasible, as will be seen later, for the entropy of the system to decrease. If this happens the entropy of the surroundings must increase by at least as much to make the overall entropy change positive.

10.3 Calculation of entropy changes

The principle of many calculations of entropy is that entropy is a property of state, and if therefore the change in entropy between initial and final states can be calculated for some process, then this is the entropy change for all processes since the entropy only depends on the initial and final states. Since $\Delta s = \int (dQ)_{rev}/T$, a reversible process is used to calculate the entropy change, and then this is the entropy change however the process actually occurs.

Example 10.1 Calculate the entropy change when a block of metal of mass m and specific heat c is heated from T_1 (K) to T_2 (K).

Solution Imagine the heating is carried out reversibly. In this case we can use:

$$\Delta S = \int_1^2 \frac{dQ_{rev}}{T} \qquad (10.6)$$

and in this case we can write that:

$$dQ_{rev} = mcdT \qquad (10.7)$$

Then substituting eqn 10.7 into eqn 10.6:

$$\Delta S = \int_1^2 mc\frac{dT}{T} \qquad (10.8)$$

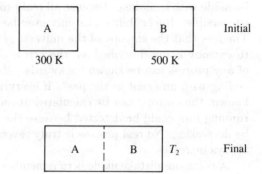

Fig. 10.1 Example 10.2: two blocks of metal brought into direct thermal contact

If the specific heat is constant and not dependent on temperature, then eqn 10.8 can be integrated:

$$S_2 - S_1 = mc \ln \frac{T_2}{T_1} \qquad (10.9)$$

and this is the entropy change however the heating from T_1 to T_2 is carried out.

Note that as the solid is heated then its entropy increases. Equation 10.9 is valid for the heating or cooling of solids or liquids which have no appreciable volume change with temperature. For this reason it is not valid for gases.

Example 10.2

Two blocks of metal, each of mass 2 kg and specific heat 0.5 kJ/kg K are brought together (see Fig. 10.1). Initially block A is at a temperature 300 K and block B is at a temperature 500 K. Calculate the final temperature of the blocks and the entropy change of each block.

Solution

The two blocks together form a closed system. They do no work on the surroundings and exchange no heat with the surroundings. Therefore from:

$$Q - W = \Delta U \qquad (10.10)$$

the internal energy is constant. Writing the internal energy of one of the metal blocks as mcT, then:

$$m_A c_A T_{1A} + m_B c_B T_{1B} = (m_A c_A + m_B c_B) T_2 \qquad (10.11)$$

or:

$$2 \times 500 \times 300 + 2 \times 500 \times 500 = (2 \times 500 + 2 \times 500)T_2$$

or:

$$T_2 = 400 \text{ K}$$

Then applying eqn 10.9 to each block:

$$\Delta S_\text{A} = 2 \times 500 \times \ln \frac{400}{300} = +288 \text{ J/K}$$

$$\Delta S_\text{B} = 2 \times 500 \times \ln \frac{400}{500} = -223 \text{ J/K}$$

and the total entropy change is:

$$\Delta S = +288 - 223 = +65 \text{ J/K}$$

The total entropy change is positive, but does this mean that the heat transfer was irreversible? This is the total entropy change **of the system**. The entropy change **of the surroundings** is in this case zero because there is no effect on, specifically no heat transfer with, the surroundings. Therefore the entropy change of the universe is positive, and consequently it can be concluded that the process was irreversible.

The source of the entropy in this example is that heat is transferred from block B to block A across a large temperature difference: initially from 500 K to 300 K. Heat transfer across a finite temperature difference is always irreversible. The system of the hot and cold blocks could be set up as a heat engine and some work produced. The next example envisages this possibility.

Example 10.3   Two blocks of metal, each of mass 2 kg and specific heat 0.5 kJ/kg K are brought to the same temperature by means of transferring heat between them and using the heat to produce work in a reversible heat engine (see Fig. 10.2). Initially block A is at a temperature 300 K and block B is at a temperature 500 K. Calculate the final temperature of the blocks and the entropy change of each block.

Solution   This problem is most easily solved by realising that there is no heat exchange with the surroundings, therefore:

$$(\Delta S)_\text{surroundings} = 0 \tag{10.12}$$

and so, since the whole process is reversible and the entropy change of the universe is zero,

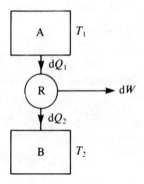

Fig. 10.2   The two blocks of metal connected by a reversible heat engine

$$(\Delta S)_{\text{system}} = 0 \tag{10.13}$$

The entropy change of the system is made up of two parts: the entropy change of block A, and the entropy change of block B. The sum of these entropy changes is zero, hence from eqn 10.9:

$$\ln \frac{T_2}{T_{1A}} + \ln \frac{T_2}{T_{1B}} = 0 \tag{10.14}$$

This gives a simple equation for the common final temperature of the blocks, T_2:

$$T_2 = \sqrt{T_{1A}T_{1B}} = \sqrt{300 \times 500} = 387.3 \text{ K}$$

in contrast to the value of 400 K obtained previously. Using eqn 10.9 the individual entropy changes can be calculated. The results are:

$$\Delta S_A = -255 \text{ J/K} \quad \text{and} \quad \Delta S_B = +255 \text{ J/K}$$

As expected the individual entropy changes are equal in magnitude and opposite in sign.

It is also simple to calculate the work produced by this heat engine system. Previously the two blocks came to a temperature of 400 K, with the heat engine they reach only 387.3 K. The energy represented by this temperature change has been produced as work, so, remembering that there are two blocks:

$$W = 2mc(400 - 387.3) = 2 \times 2 \times 500 \times 12.7 = 25400 \text{ J}$$

Before entropy changes in an ideal gas can be calculated it is necessary to consider a way of combining the first and second laws which leads to some of the most useful equations in thermodynamics.

10.4 The combined first and second laws

The first law of thermodynamics for closed systems can be expressed for a differential change on unit mass of any substance as:

$$dQ - dW = du \tag{10.15}$$

For a reversible change in a closed system $dW = pdv$, and then using the second law the reversible heat transfer can be written as $dQ = Tds$. Making these substitutions in eqn 10.15 gives:

$$Tds - pdv = du \tag{10.16}$$

Although eqn 10.16 was derived from eqn 10.15 using identities only true in a reversible process, eqn 10.16 is true for any kind of process. This is because eqn 10.16 contains only properties of state and so the equation is not dependent on a particular path being taken. Equation 10.16 is so useful because it is always true for any substance and for any type of change.

An even more useful equation can be derived by using the differential form of $h = u + pv$, that is:

$$dh = du + pdv + vdp \tag{10.17}$$

and eliminating du between eqns 10.16 and 10.17:

$$dh = Tds + vdp \tag{10.18}$$

This equation, like eqn 10.16, is true for any substance and for any kind or process.

A number of simple illustrations of the use of this equation are now given.

1. Isentropic compression of a liquid, for example in the feed pump of a Rankine cycle. Using eqn 10.18, for the case of constant entropy:

$$dh = vdp \tag{10.19}$$

and so:

$$h_2 - h_1 = \int_1^2 vdp = v(p_2 - p_1) \tag{10.20}$$

as derived previously, if again it can be assumed that the specific volume of water is not dependent upon the pressure.

2. Derivation of the shaft work in a reversible steady-flow process. From the steady-flow energy equation per unit mass of substance flowing:

$$dQ - dW_s = dh \qquad (10.21)$$

and substituting for dh from eqn 10.18, gives:

$$dQ - dW_s = Tds + vdp \qquad (10.22)$$

If the process is reversible, then dQ and Tds can be equated in eqn 10.22. Integration then gives the result for the shaft work W_s:

$$dW_s = - \int vdp \qquad (10.23)$$

10.5   Entropy changes in an ideal gas

Explicit equations can de derived for the dependence of entropy in an ideal gas on pressure, temperature, and volume.

The starting point for the first of these derivations is eqn 10.16 together with the fact that since internal energy, for an ideal gas, is a function of temperature only then:

$$du = c_v dT \qquad (10.24)$$

Substituting this equation into eqn 10.16 gives:

$$c_v dT = Tds - pdv \qquad (10.25)$$

or:

$$ds = c_v \frac{dT}{T} + \frac{p}{T} dv \qquad (10.26)$$

Then, because we are interested in unit mass of an ideal gas substitute R/v for p/T to give:

$$ds = c_v \frac{dT}{T} + R \frac{dv}{v} \qquad (10.27)$$

Then, if c_v can be treated as a constant, this equation can be integrated to produce the desired result:

$$s_2 - s_1 = c_v \ln \frac{T_2}{T_1} + R \ln \frac{v_2}{v_1} \qquad (10.28)$$

Here we have the entropy change of an ideal gas related to changes in temperature and volume. Equation 10.28 is true for any type of change in an ideal gas: it is not restricted to reversible changes. It expresses the entropy change as a function of the change of two other properties of state: temperature and volume.

By using eqn 10.18 and $dh = c_p dT$ the entropy change may be expressed in terms of T and p:

$$s_2 - s_1 = c_p \ln \frac{T_2}{T_1} - R \ln \frac{p_2}{p_1} \qquad (10.29)$$

Finally, for example from eqn 10.29, by eliminating the temperature using:

$$\frac{T_2}{T_1} = \frac{p_2}{p_1} \times \frac{v_2}{v_1} \qquad (10.30)$$

and remembering that for an ideal gas $c_p - c_v = R$, gives:

$$s_2 - s_1 = c_p \ln \frac{v_2}{v_1} + c_v \ln \frac{p_2}{p_1} \qquad (10.31)$$

Equations 10.28, 10.29, and 10.31 all relate entropy changes to different sets of two properties of state, and all are true for ideal gases undergoing all types of processes.

Example 10.4 Find the equation relating p and v in an isentropic change in an ideal gas.

Solution Since the required equation is to contain p and v, eqn 10.31 is the logical place to start. For constant entropy this becomes:

$$c_p \ln \frac{v_2}{v_1} + c_v \ln \frac{p_2}{p_1} = 0 \qquad (10.32)$$

or:

$$c_p \ln v + c_v \ln p = C_1 \qquad (10.33)$$

where C_1 is a constant. Defining $\gamma = c_p/c_v$, eqn 10.33 implies that:

$$pv^\gamma = C_2 \qquad (10.34)$$

where C_2 is a constant. This is the equation for an isentropic change in an ideal gas. The ideal gas equation can be used to express the equation in terms of p and T:

$$\frac{T^\gamma}{p^{\gamma-1}} = C_3 \qquad (10.35)$$

where C_3 is another constant.

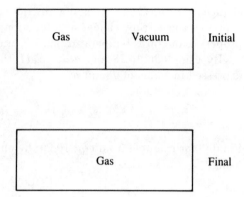

Fig. 10.3   Example 10.5: an ideal gas expands into a vacuum

For air c_p=1005 J/kg K and c_v=718 J/kg K, thus γ=1.40. The values of γ for other gases are discussed in Chapter 11.

Example 10.5
A rigid, thermally insulated box contains two compartments of equal volume (see Fig. 10.3). Initially the left-hand compartment contains 1 kilogram-mole of ideal gas, and the right-hand compartment contains a vacuum. The partition is ruptured: calculate the entropy change of the gas.

Solution
Assume the whole box is a control volume for the closed-system version of the first law $Q - W = \Delta U$. In this case $Q = 0$ and $W = 0$, so the internal energy is constant. Since the gas is ideal, the temperature is constant. So the process is that the gas expands from $V_1 = V$ to $V_2 = 2V$ at constant temperature. The relevant equation for the entropy change is therefore the one containing temperature and volume:

$$s_2 - s_1 = c_v \ln \frac{T_2}{T_1} + R \ln \frac{v_2}{v_1} \qquad (10.36)$$

Putting $T_2/T_1 = 1$ and $V_2/V_1 = 2$, then the entropy change per unit mass is given by:

$$s_2 - s_1 = R \ln 2$$

and so the entropy change for a kilogram-mole is:

$$S_2 - S_1 = R_0 \ln 2 = 8314 \ln 2 = +5763 \text{ J/K}$$

Note that the entropy change of the system is positive. The entropy change of the surroundings is zero as there is no heat exchange with the surroundings. Hence the overall entropy change of the universe was positive, and so the process was actually irreversible.

Example 10.6

A previous example of an adiabatic nozzle air flow had the following initial and final conditions:

	temperature (K)	pressure (bar)	velocity (m/s)
initial condition	298	2	zero
final condition	255	1	294

Is this a reversible flow?

Solution

The velocity is not relevant. The entropy change of the surroundings is zero since the flow is adiabatic. Hence the reversibility can be judged on the entropy change of the air alone. Since pressure and temperature are quoted the relevant equation is:

$$s_2 - s_1 = c_p \ln \frac{T_2}{T_1} - R \ln \frac{p_2}{p_1} \qquad (10.37)$$

Substituting into this equation for this air flow:

$$s_2 - s_1 = 1005 \ln \frac{255}{298} - 287 \ln \frac{1}{2} = +42 \text{ J/kg K}$$

Hence the overall entropy change is positive, and the flow is irreversible.

10.6 Entropy changes in steam–water systems

For steam–water systems simple equations for entropy changes do not apply, and as for internal energy and enthalpy, values for the entropy are tabulated. As for these other properties, there is commonly a saturation table and a superheated steam table. Table 10.1 shows part of the saturation table. Here, as for internal energy and enthalpy, a subscript f indicates saturated water, and a subscript g indicates saturated steam. Hence:

s_f = entropy per kilogram of saturated water (J/kg K)
s_g = entropy per kilogram of saturated steam (J/kg K)
s_{fg} = $s_g - s_f$, the entropy of vaporization per kg (J/kg K)

The superheated steam table, part of which is shown in Table 10.2, gives the entropy, and other properties, as function of temperature and pressure. Note that in the two–phase region, where the phases are saturated, the entropy of each phase depends only on temperature or

Table 10.1 Values of h, and s for saturated water and steam, to 221 bar

p (bar)	t_s (°C)	h_f (kJ/kg)	h_{fg} (kJ/kg)	h_g (kJ/kg)	s_f (kJ/kg K)	s_{fg} (kJ/kg K)	s_g (kJ/kg K)
0.00611	0.01	0.0	2501.6	2501.6	0.0	9.1575	9.1575
0.0233	20.00	83.9	2454.3	2538.2	0.2963	8.3721	8.6684
1.00	99.63	417.5	2257.9	2675.4	1.3027	6.0571	7.3598
10.00	179.88	762.6	2013.6	2776.2	2.1382	4.4446	6.5828
100.00	310.96	1408.1	1319.6	2727.7	3.3606	2.2592	5.6198
221.20	374.15	2107.4	0.0	2107.4	4.4429	0.0	4.4429

Table 10.2 Values of v, u, h, and s for superheated steam, to 220 bar and 500 °C,

p (bar) (t_s (°C))		Temperature (°C)					
		t_s	100	200	300	400	500
1.00	v (dm³/kg)	1693.7	1695.5	2172.3	2638.7	3102.5	3565.3
(99.60)	u (kJ/kg)	2506.0	2506.7	2658.2	2810.6	2968.0	3131.6
	h (kJ/kg)	2675.4	2676.2	2875.4	3074.5	3278.2	3488.1
	s (kJ/kg K)	7.3598	7.3618	7.8349	8.2166	8.5442	8.8348
10	v (dm³/kg)	194.30		205.92	257.98	306.49	353.96
(179.9)	u (kJ/kg)	2581.9		2620.9	2794.1	2957.9	3124.3
	h (kJ/kg)	2776.2		2826.8	3052.1	3264.4	3478.3
	s (kJ/kg K)	6.5828		6.6922	7.1251	7.4665	7.7627
100	v (dm³/kg)	18.041				26.41	32.76
(311.0)	u (kJ/kg)	2547.3				2835.8	3047.0
	h (kJ/kg)	2727.7				3099.9	3374.6
	s (kJ/kg K)	5.6198				6.2182	6.5994

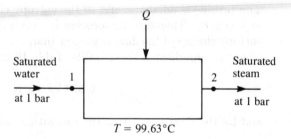

Fig. 10.4 Example 10.7: boiler converts saturated water into saturated steam

pressure. In the single–phase superheated region, the steam entropy is a function of two properties of state: here pressure and temperature.

The use of the tables is illustrated by two simple examples.

Example 10.7

Saturated water at 1 bar flows steadily through a boiler, and emerges as dry saturated steam at 1 bar (see Fig. 10.4). Calculate, from the enthalpy change, the change in entropy per kilogram flowing.

Solution

From the steady-flow energy equation, per unit mass ignoring the kinetic and potential energy terms:

$$Q - W_s = h_2 - h_1 = h_g - h_f = h_{fg}$$

Here there is no shaft work, and so substituting for h_{fg} from Table 10.1:

$$Q = h_{fg} = 2257.9 \text{ kJ/kg}$$

This heat transfer takes place at constant temperature, the saturation temperature at 1 bar, which from Table 10.1 is 99.63°C or 372.78 K. Since the heat transfer takes place at this constant temperature:

$$s_2 - s_1 = \int \frac{dQ}{T} = \frac{Q}{T} = \frac{2257.9}{372.78} = 6.057 \text{ kJ/kg K}$$

This value is the entropy of vaporization, s_{fg} tabulated in Table 10.1.

Example 10.8

The earlier example of a Rankine cycle (example 7.3) had an adiabatic turbine where the inlet was steam at 500°C and 100 bar. The outlet was steam at 20°C and a dryness (x) of 0.8. Calculate the entropy change, and show that the turbine is irreversible.

Solution   The turbine is adiabatic and so the entropy change of the surroundings is zero. Therefore the process is irreversible in this case, if the
entropy change of the fluid is greater than zero. First the initial state,
from the superheated steam table, Table 10.2:

$$s_1 = 6.599 \text{ kJ/kg K}$$

and for the final state, from the saturation table, Table 10.1:

$$
\begin{aligned}
s_2 &= (1 - x)s_\mathrm{f} + xs_\mathrm{g} = s_\mathrm{f} + xs_\mathrm{fg} \\
&= 0.2963 + 0.8 \times 8.3721 = 6.994 \text{ kJ/kg K}
\end{aligned}
$$

So:

$$(\Delta s)_\text{overall} = (\Delta s)_\text{system} = s_2 - s_1 = +0.395 \text{ kJ/kg K}$$

Because the overall entropy change is positive, the turbine is irreversible.

10.7   Problems

10.1 Calculate the entropy changes for the following processes

(a) A block of metal ($m = 2$ kg, $C = 400$ J/kg K) is cooled from
$100°$C to $20°$C.

(b) Liquid water ($m = 1$ kg, $C = 4200$ J/kg K) is cooled from $100°$C
to $20°$C. Compare the calculated value with the result from steam
tables.

(c) An ideal gas ($m = 1$ kg, properties of air) is cooled from $100°$C
to $20°$C at constant pressure.

(d) An ideal gas ($m = 1$ kg, properties of air) is cooled from $100°$C
to $20°$C at constant volume.

(e) Liquid water ($m = 5$ kg, $C_\text{water} = 4200$ J/kg K) at $20°$C is cooled
until it freezes, then the ice is cooled to -$20°$C. Latent heat of
freezing $= 333$ kJ/kg, $C_\text{ice} = 2100$ J/kg K.

10.2 Calculate the entropy change of the system for the following cases. Is
the change reversible or irreversible?

(a) The thermally insulated, rigid vessel in question 6.1.

(b) The thermally insulated, rigid vessel in question 6.8.

(c) The adiabatic air flow in question 7.1.

(d) The adiabatic air turbine in question 7.2.

(e) The adiabatic steam turbine in question 7.5.

What is it about these processes that allows a conclusion about the reversibility of the process?

10.3 Three blocks of metal have the same mass and specific heat. Their initial temperatures are T, $2T$, and $3T$. Reversible heat engines and heat pumps are connected between these blocks, but no net work is produced or used. No heat is exchanged with the surroundings. Finally two of the blocks reach the same temperature T_1, and the third block reaches T_2. Find these temperatures. [Hint. Write down the first law and the second law in forms only involving the initial and final temperatures.]

10.4 A compressor takes in air at 1 bar, 300 K at a steady flow of 5 kg/s and delivers it to a large vessel at 6 bar. Calculate the power required if the compressor is reversible and (a) adiabatic, (b) isothermal. In case (a) calculate also the air temperature after compression. Sketch the processes on a p–v diagram.

10.5 Air from a large vessel at 400 K, 2 bar flows steadily with negligible friction through an adiabatic nozzle to atmospheric pressure at 1 bar. Calculate the air temperature, the velocity at exit and the change in specific entropy.

10.6 Air from a large vessel at 400 K, 2 bar flows steadily controlled by viscous friction down a thermally insulated capillary tube to the atmosphere at 1 bar. Calculate the air temperature, and the change in specific entropy.

10.7 Air flows along a thermally insulated pipe of varying cross section. At position X the temperature and pressure are 500 K and 4 bar. At position Y they are 600 K and 6 bar. Which way is the air flowing?

10.8 Calculate the work produced when 1 kg of steam at 200°C and 4 bar is expanded adiabatically and reversibly to: (a) 3 bar, (b) 2 bar, (c) 1 bar.

11

Molecular interpretation of thermodynamic properties

11.1 Key points of this chapter

This chapter considers how some thermodynamic properties arise from molecular considerations. Readers particularly interested in this topic should consult a specialized textbook on statistical thermodynamics.

- The main concepts of the kinetic theory of gases are explained. The gas molecules are in continuous motion, and this gives rise to the pressure on a surface, and to transport phenomena like viscosity and thermal conductivity. (Section 11.2)

- The main results from the kinetic theory of gases are given: for the velocity distribution, the pressure, the flux of molecules at a surface, and the transport properties. (Section 11.2)

- Internal energy arises from the different types of energy possessed by the molecules: kinetic energy of translation and rotation, and vibrational energy. (Section 11.3)

- The principle of the equipartition of energy shows how the energy is shared between these various types of energy. (Section 11.3)

- Equations for the internal energy can be interpreted in the form of specific heats: both at constant volume and constant pressure. (Section 11.3)

- From the values of the two principal specific heats, their ratio γ can be calculated. The calculated values agree well with experimental values. (Section 11.4)

• Entropy is related to the probability that a state will occur. Disorder is more likely to occur than an ordered system. Transition to a higher probability, more dis-ordered state involves an increase in entropy. (Section 11.5)

11.2   The kinetic theory of gases

The kinetic theory of gases developed from the idea that gas molecules are continually in motion, and that this motion can explain many phenomena occurring in gases as follows.

1. When the gas molecules hit a surface and rebound, their momentum changes. To provide this change in momentum there must be a force. This force is the origin of the pressure exerted by the gas on the walls of the containing vessel.

2. The gas molecules are in continuous and random motion. A particular molecule moves in straight lines in between collisions with other molecules and with the vessel walls. A molecule therefore has a finite probability of moving from one place to another. This is the origin of diffusion. If the stopper is removed from a bottle of volatile liquid, the odour of the liquid gradually pervades the air in even a draught-free room.

3. The gas molecules have a distribution of velocities about a mean. The mean speed is related to the temperature: the higher the temperature the higher the mean speed. As the gas molecules in a region of high gas temperature gradually diffuse to a cooler region, they carry with them their relatively high velocity. This is the origin of molecular thermal conduction in gases.

4. The molecules not only have a high random velocity about a mean velocity of zero[1], but also have a bulk velocity due to any imposed flow. As the gas molecules diffuse from a region of high bulk velocity to a region of low bulk velocity, they carry with them the high bulk velocity. As the high velocity is reduced to the lower bulk velocity of the molecules around them a shear force is produced. This is the origin of molecular viscosity.

---

[1]Velocity is a vector, so taking into account the direction the mean velocity in a stagnant gas is zero. If the word 'speed' is used for the magnitude of the velocity vector, the average speed is certainly not zero.

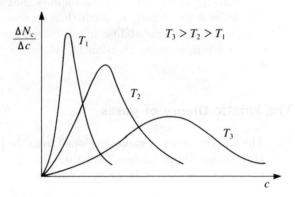

Fig. 11.1   The Maxwell–Boltzmann velocity distribution

It is not difficult to produce drastically over-simplified versions of many of the derivations of the main results of the kinetic theory of gases, but much more difficult to produce really convincing derivations. The problem is that there is always a distribution of gas velocities, and the molecules are moving in random directions. The mathematics is complicated by many multiple integrals as integrations over all velocities and all directions are carried out. Here only the main results of the kinetic theory of gases will be quoted.

1. The velocity of the gas molecules follows the well-known Maxwell–Boltzmann distribution (see Fig. 11.1);

$$\frac{\Delta N_c}{N} = \frac{4c^2 \Delta c}{\sqrt{\pi} c_m^3} \exp\left(\frac{-c^2}{c_m^2}\right) \tag{11.1}$$

where ΔN_c is the number of molecules with speeds between c and $c + \Delta c$, N is the total number of molecules, and c_m is the most probable speed of the molecules, see eqn 11.2 below. Figure 11.1 shows the distribution at three different temperatures. Note that as the temperature increases the distribution gets broader as well as shifting to higher speeds.

2. For this velocity distribution there are three significant speeds (the term speed is used since we are not concerned with the direction of the velocity), which are:

(a) the most probable speed, c_m, which is the peak of the distribution in Fig. 11.1, defined by:

$$c_m = \sqrt{2RT} \tag{11.2}$$

(b) the mean speed, $\bar{c}$, defined by:

$$\bar{c} = \sqrt{\frac{8RT}{\pi}} \tag{11.3}$$

(c) the root mean square speed, c_{RMS}, is the square root of the mean of the square of all the velocities, defined by:

$$c_{RMS} = \sqrt{3RT} \tag{11.4}$$

These three velocities are in the ratio:

$$c_m : \bar{c} : c_{RMS} = 1.000 : 1.128 : 1.224 \tag{11.5}$$

3. The pressure exerted by the gas at a surface is given by:

$$p = \frac{mn\overline{c^2}}{3} \tag{11.6}$$

where m is the mass of each molecule (kg), n is the number of molecules per cubic meter, and $\overline{c^2}$ is the mean square speed $= c_{RMS}^2$. Substituting for c_{RMS} from eqn 11.4 the pressure can be expressed as:

$$p = mnRT = RT/v \tag{11.7}$$

because mn is the mass of gas per unit volume, that is the density or $1/v$. Equation 11.7 is, of course, the ideal gas equation.

4. The flux of molecules hitting a surface from one side (the number per unit area per unit time) is given by $n\bar{c}/4$.

5. The mean free path, λ, for the molecules is the average distance a molecule travels between collisions with other molecules, and is given by:

$$\lambda = \frac{1}{\sigma n} \tag{11.8}$$

σ is the collision cross section of the molecule or $\pi(2\rho)^2$ where ρ is the molecular radius. Note that any molecule whose centre comes within a distance 2ρ of the centre of any other molecule will experience a collision.

6. The molecular viscosity, μ (kg/m s) is given by:

$$\mu = \frac{m\bar{c}}{3\sigma} \tag{11.9}$$

7. Similarly the molecular thermal conductivity, κ (W/m K) is given by:

$$\kappa = \frac{m\bar{c}c_v}{3\sigma} \tag{11.10}$$

Note that it is a prediction of the kinetic theory of gases from eqns 11.9 and 11.10 that:

$$\frac{\mu c_v}{\kappa} = 1 \tag{11.11}$$

The following example illustrates the use of these equations.

Example 11.1

For nitrogen at 300 K and 1 bar calculate values of c_m, $\bar{c}$, c_{RMS}, λ, μ, and κ. Use a molecular radius (ρ) of 1.83×10^{-10} m. For nitrogen $R = 8314/28 = 297$ J/kg K, and $c_v = 718$ J/kg K.

Solution

Using equations from earlier:

1. From eqn 11.2:

$$c_m = \sqrt{2RT} = \sqrt{2 \times 297 \times 300} = 422 \text{ m/s}$$

2. From eqn 11.3:

$$\bar{c} = \sqrt{\frac{8RT}{\pi}} = \sqrt{\frac{8 \times 297 \times 300}{\pi}} = 476 \text{ m/s}$$

3. From eqn 11.4:

$$c_{RMS} = \sqrt{3RT} = \sqrt{3 \times 297 \times 300} = 517 \text{ m/s}$$

4. To calculate the mean free path first calculate n, the number of molecules per unit volume. The molar volume (the volume occupied by 1 kilogram-mole), V_{molar} is:

$$V_{molar} = \frac{R_0 T}{p} = \frac{8314 \times 300}{10^5} = 24.94 \text{m}^3$$

This volume contains an Avogadro's number of molecules (6.023×10^{26}), so:

$$n = \frac{6.023 \times 10^{26}}{24.94} = 2.41 \times 10^{25} \text{molecules/m}^3$$

The collision cross section σ is:

$$\sigma = \pi(2\rho)^2 = \pi(2 \times 1.83 \times 10^{-10})^2 = 4.21 \times 10^{-19} \text{m}^2$$

and then from eqn 11.8:

$$\lambda = \frac{1}{\sigma n} = \frac{1}{4.21 \times 10^{-19} \times 2.41 \times 10^{25}} = 9.84 \times 10^{-8} \text{m}$$

5. The mass of an individual molecule m is the relative molar mass (28) divided by Avogadro's number (6.023×10^{26}), that is 4.65×10^{-26} kg. Then from eqn 11.9:

$$\mu = \frac{m\bar{c}}{3\sigma} = \frac{4.65 \times 10^{-26} \times 476}{3 \times 4.21 \times 10^{-19}} = 17.5 \times 10^{-6} \text{ Ns/m}^2$$

6. From eqn 11.10:

$$\kappa = \frac{m\bar{c}c_v}{3\sigma} = \frac{4.65 \times 10^{-26} \times 476 \times 718}{3 \times 4.21 \times 10^{-19}} = 0.013 \text{ W/m K}$$

The actual values of the viscosity and the thermal conductivity are:

$$\mu = 18 \times 10^{-6} \text{ Ns/m}^2 \quad \text{and} \quad \kappa = 0.026 \text{ W/m K}$$

The accuracy of the viscosity is very good, but the thermal conductivity is not nearly so good. However, here the molecular radius has been deduced from actual viscosity values! More complex theories and equations can give better values for the thermal conductivity.

Another test of the simple kinetic theory is that the value of $\mu c_v/\kappa$, which according to eqn 11.11 should be 1, is actually around 0.5. An example of the value of the kinetic theory is that it is not difficult to show that μ should be proportional to $T^{-\frac{1}{2}}$, and such knowledge helps greatly in the interpolation and extrapolation of tables of properties.

11.3 Internal energy and specific heats

Previously it has been said that the internal energy is the energy of the molecules, but what kind of energy? If the molecules are imagined as discrete, very small atoms joined by bonds then various types of energy are possible.

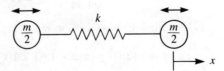

Fig. 11.2   Vibrating spring system

1. The molecules can have kinetic energy due to the the motion of the centre of mass: this is kinetic energy of translation. For a mass m with a velocity of the centre of mass c, the kinetic energy E_t is:

$$E_t = \frac{1}{2}mc^2 \tag{11.12}$$

2. The molecules can have kinetic energy due to a rotation about the centre of mass: this is kinetic energy of rotation. For a body with moment of inertia I about the centre of mass and an angular velocity ω about the centre of mass, the kinetic energy E_r is:

$$E_r = \frac{1}{2}I\omega^2 \tag{11.13}$$

3. The molecules can also have vibrational energy. If the molecule is imagined as two masses joined by a spring, then when the system vibrates (see Fig. 11.2) the energy of the system is partly potential energy (energy stored in the compressed spring), and partly kinetic energy (due to the motion of the two masses). As the system vibrates potential energy is converted into kinetic energy and vice versa. The potential energy E_{sp} of a compressed or extended spring is:

$$E_{sp} = \frac{1}{2}kx^2 \tag{11.14}$$

where k is the force in the spring per unit extension (N/m), and x is the spring extension from the rest position (m). Thus the total energy of the vibrating system E_v is:

$$E_v = \frac{1}{2}kx^2 + \frac{1}{2}m\dot{x}^2 \tag{11.15}$$

where here m is the total mass of the vibrating system in Fig. 11.2.

It should be noted that each of the different types of energy in eqns 11.12, 11.13, and 11.15 contains squared terms: c^2, ω^2, x^2, or $\dot{x}^2$.

One of the fundamental results of statistical thermodynamics is the principle of the **equipartition of energy**, which can be stated in the following simple form. 'The energy to be expected for any part of the total energy which can be expressed as a sum of squares is of amount $\frac{1}{2}RT$ for every squared term in this part of the energy.'

Take first a molecule which consists of a single atom, for example helium. There is no possibility of vibration. If we also make the assumption that the helium atom is essentially a point mass of almost zero size, then the moment of inertia is very small and there is no significant kinetic energy of rotation. The velocity c of the atoms can be expressed as:

$$c^2 = c_x^2 + c_y^2 + c_z^2 \tag{11.16}$$

where c_x is the velocity in the x-axis direction etc. Thus there are here three squared terms in the total energy of the helium atom. Hence from the equipartition principle, the total energy (the internal energy) of the helium is:

$$u = \frac{3}{2}RT \tag{11.17}$$

and so the specific heat at constant volume c_v, being du/dT for an ideal gas, is:

$$c_v = \frac{3}{2}R \tag{11.18}$$

and since for an ideal gas the specific heat at constant pressure, c_p, is given by $c_v + R$:

$$c_p = \frac{5}{2}R \tag{11.19}$$

Substituting values into eqn 11.19 for helium:

$$c_p = \frac{5}{2}R = \frac{5 \times 8314}{2 \times 4} = 5200 \text{ J/kg K}$$

The experimental value is 5190 J/kg K, and so the theory works very well. Equations 11.17, 11.18, and 11.19 are valid for all monatomic gas molecules.

Now consider diatomic molecules, such as hydrogen, oxygen, and nitrogen. These can be imagined as being like the spring system in Fig. 11.2. These molecules can move in three directions like the single atom, but they also have rotational kinetic energy about two axes. There is no significant kinetic energy of rotation about the third axis along the line joining the atoms because about this the moment of

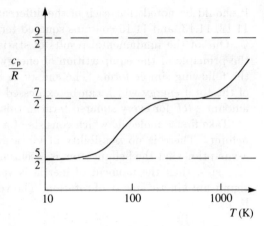

Fig. 11.3 Variation of c_p for hydrogen with temperature

inertia is extremely small. There are thus five squared terms in the total energy of the a diatomic molecule. Hence from the equipartition principle, the total energy (the internal energy) is:

$$u = \frac{5}{2}RT \tag{11.20}$$

Then, as before for the helium example:

$$c_v = \frac{5}{2}R \tag{11.21}$$

and:

$$c_p = \frac{7}{2}R \tag{11.22}$$

Substituting values into eqn 11.22 for hydrogen:

$$c_p = \frac{7}{2}R = \frac{7 \times 8314}{2 \times 2} = 14600 \text{ J/kg K}$$

The experimental value is 14200 J/kg K, and so the theory works very well. Equations 11.20, 11.21, and 11.22 work well for many monatomic gas molecules. There is, however, a difficulty. A diatomic molecule ought to be able to vibrate and therefore contribute another two squared terms, thus making $c_p = \frac{9}{2}R$. However, from the experimental results, this does not appear to be the case. The answer to this difficulty lies in quantum mechanics. Figure 11.3 shows some experimental results for the variation of c_p with temperature for hydrogen. This figure shows the following.

1. Below about 50 K, $c_p = \frac{5}{2}R$, and the result is consistent with no rotation or vibration occurring.

2. Around 300 K, $c_p = \frac{7}{2}R$, and the result is consistent with no vibration occurring.

3. Above about 700 K, the c_p begins to rise to around $\frac{9}{2}R$ consistent with translation, rotation, and vibration all occurring, but the results begin to be contaminated with the effects of the hydrogen molecule splitting up into hydrogen atoms at high temperatures by the dissociation reaction:

$$H_2 \rightleftharpoons H + H$$

Quantum mechanics provides the answers as to why various types of motion are activated at different temperatures, but this is outside the scope of this book.

More complicated molecules such as water (H_2O) and methane (CH_4) again do not vibrate freely at ordinary temperatures, but they do move in three directions, and also have kinetic energy of rotation about three axes. This is because for the case of water, the molecule is bent about the central oxygen atom, and for the case of methane, the molecule is three-dimensional[2]. Thus we should have:

$$u = \frac{6}{2}RT \tag{11.23}$$

Then, as before for the helium example:

$$c_v = \frac{6}{2}R \tag{11.24}$$

and:

$$c_p = \frac{8}{2}R \tag{11.25}$$

Substituting values into eqn 11.25 for water vapour:

$$c_p = \frac{8}{2}R = \frac{8 \times 8314}{2 \times 18} = 1847 \text{ J/kg K}$$

The experimental value is 1890 J/kg K. There is a tendency for the theory to work less well for complicated molecules where the results, even at room temperature, are contaminated by the effects of vibration.

---

[2]In fact the four hydrogen atoms are arranged at the vertices of a regular tetrahedron.

Table 11.1   Values of γ for various gases

Gas	Type of molecule	$\gamma = c_p/c_v$ experimental	$\gamma = c_p/c_v$ theoretical
helium	monatomic	1.67	1.67
neon	monatomic	1.67	1.67
argon	monatomic	1.67	1.67
hydrogen	diatomic	1.41	1.40
oxygen	diatomic	1.40	1.40
nitrogen	diatomic	1.40	1.40
water	triatomic	1.32	1.33
methane (CH_4)	5 atoms	1.30	1.33

11.4   The ratio of the specific heats for gases

From the equations in the last section it is now easy to produce a theoretical value for γ, the ratio of the specific heat at constant pressure to the specific heat at constant volume. Table 11.1 shows the predictions for γ and the actual values for a number of different gases. In all cases it has been assumed that vibration in the molecule does not occur. This is not true for some of the molecules, and explains why all the theoretical values for the more complicated molecules are too high.

11.5   Entropy

It is easy to visualize the meaning of internal energy on a molecular scale: it represents the energy, kinetic energy of translation and rotation, and vibrational energy, of the molecules. Entropy is much more difficult to visualize in any clear sense. The most meaningful way to visualize entropy is to relate it to the probability that various states occur. Imagine a closed box containing some gas molecules. The box is divided into two compartments, but the partition has a hole so that molecules can move from one compartment to another. If the probability of finding any particular molecule in the left-hand compartment is $\frac{1}{2}$, then if there are n molecules in total the probability of finding all of them in the left hand compartment is $(\frac{1}{2})^n$. If n is large, and even a small box can contain 10^{20} molecules, then this probability of finding all these molecules in one compartment is exceedingly small.

Boltzmann first realised the connection between the probability

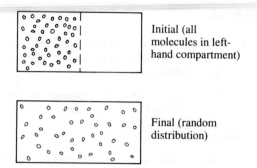

Fig. 11.4   Statistical interpretation of entropy

that a state occurs and the entropy. He first used the equation for the entropy per molecule:

$$s = k \ln P \tag{11.26}$$

where k is Boltzmann's constant $= 1.380 \times 10^{-23}$ J/K, and P is the probability of a state occurring. Boltzmann's constant can be thought of as the universal gas constant per molecule, and is therefore related to R_0 and Avogadro's number, N:

$$k = \frac{R_0}{N} = \frac{8314}{6.023 \times 10^{26}} = 1.380 \times 10^{-23} \text{ J/K} \tag{11.27}$$

Hence from eqn 11.26 the more likely a state is the higher will be its entropy. If all the gas starts of in the left-hand compartment by means of a solid wall separating the two compartments, and then the wall is ruptured, this means that the entropy of the system will rise as the gas spreads between the two compartments. As was found in Chapter 10 the entropy does indeed rise, and this rise was previously ascribed to the drop in pressure as the gas expands.

An alternative, but related, view is to think of the connection between entropy and order or disorder. The gas in the left-hand compartment only in Fig. 11.4 can be thought of as an ordered situation, and the gas dispersed between both compartments as a less ordered situation. Less order and increasing entropy are inextricably linked. Similarly the melting of ice, a relatively ordered crystal structure, to form the disordered water structure is also accompanied by an increase in entropy.

11.6   Problems

11.1 From eqn 11.1 prove that c_m is indeed the most probable speed, and that the maxima in Fig. 11.1 are greater at low temperatures than at high temperatures.

11.2 Estimate c_p and c_v for chlorine gas: for basic data see question 2.1. Chlorine exists as a diatomic gas Cl_2.

11.3 Using Fig. 11.3 estimate c_p and c_v for hydrogen at room temperature.

11.4 Why is helium sometimes used for cooling high-temperature nuclear reactors?

11.5 From Fig. 11.3 estimate the value of c_p for hydrogen at very high temperatures. Is any higher specific heat possible for any substance? If a higher specific heat was apparently measured how could it be explained?

11.6 A box containing gas is divided into two compartments but the dividing wall contains a small hole. Compartment A is maintained at 300 K, and compartment B at 200 K. Calculate the ratio of the pressures p_A/p_B: (a) when the hole is large; (b) when the hole is small compared with the mean free path of the gas molecules.

12

Property diagrams

12.1 Key points of this chapter

- It is convenient to represent many thermodynamic cycles on a two-dimensional map. The co-ordinates of this map are two properties of state. Since in general two properties of state are necessary to define the state of a particular system, this two-dimensional map can be used to describe various cycles. Here the principal maps, or property diagrams, are described, starting with the simplest p–v and T–s diagrams for ideal gases. (Section 12.2)

- In a reversible closed-system cycle, the area under a p–v curve is the work, and the area under a T–s diagram is the heat transfer. (Section 12.2)

- For an ideal gas the shapes of various common processes are defined. (Section 12.2)

- For a vapour–liquid system the p–v and T–s diagrams have characteristic shapes, though the superheated vapour behaves like an ideal gas. (Section 12.3)

- A temperature–entropy diagram can reveal where a cycle falls short of the Carnot ideal, and can help in the calculation of efficiency. (Section 12.3)

- A diagram of enthalpy against entropy (a Mollier diagram) is useful in practical situations for calculations involving superheated and wet vapour. A Mollier diagram is commonly used for steam calculations. (Section 12.4)

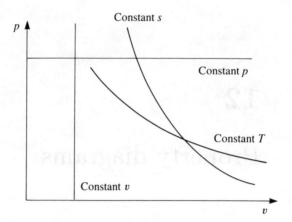

Fig. 12.1 p–v diagram for an ideal gas

- In a similar way a diagram of pressure against enthalpy is often useful in refrigeration calculations. (Section 12.5)

12.2   Pressure–volume and temperature–entropy diagrams for ideal gases

Figure 12.1 shows a pressure–volume diagram for an ideal gas. Conventionally pressure is plotted vertically and volume horizontally. As explained previously in Chapter 6, the area under a p–v curve represents the reversible work W_{rev} per unit mass in a closed system, since:

$$W_{rev} = \int p\,dv \tag{12.1}$$

Shown on Fig. 12.1 are four lines for different types of processes.

1. Obviously an isobaric (constant pressure) process is a straight horizontal line.

2. Similarly an isochoric (constant volume) process is a straight vertical line.

3. An isothermal (constant temperature) process is for an ideal gas a rectangular hyperbola:

$$pv = C \tag{12.2}$$

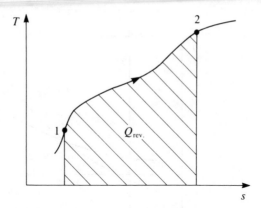

Fig. 12.2   Reversible heat transfer and the *T–s* diagram

4. An isentropic (constant entropy) process is rather similar to the isothermal line but is steeper. For an ideal gas this line is:

$$pv^\gamma = C \tag{12.3}$$

The heat transfer in a reversible process per unit mass is given by:

$$Q_{\text{rev}} = \int T ds \tag{12.4}$$

and so this heat transfer is represented by the area under a curve on a graph of entropy (plotted horizontally) and temperature (plotted vertically), see Fig. 12.2. Figure 12.3 shows isobaric, isochoric, isothermal, and isentropic process on a *T–s* diagram for an ideal gas. The isothermal and isentropic lines are obviously straight horizontal and vertical lines. The constant pressure and constant volume lines are more complicated.

The relationship between entropy, pressure, and temperature for an ideal gas is:

$$s = c_v \ln T + R \ln v \tag{12.5}$$

and so:

$$\left(\frac{\partial T}{\partial s} \right)_v = \frac{T}{c_v} \tag{12.6}$$

Therefore the slope of a line of constant volume on a *T–s* diagram is T/c_v. This means that the slope is positive, and at higher temperatures has an ever increasing slope, as shown in Fig. 12.3.

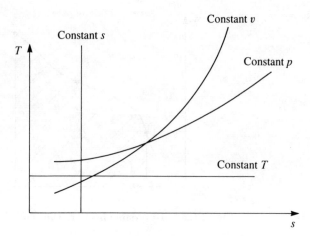

Fig. 12.3 *T–s* diagram for an ideal gas

For a constant pressure process the corresponding equations are:

$$s = c_p \ln T - R \ln p \qquad (12.7)$$

$$\left(\frac{\partial T}{\partial s}\right)_p = \frac{T}{c_p} \qquad (12.8)$$

so that the slope of the constant pressure line is T/c_p.

It should be noted that since necessarily for a gas which expands on heating $c_p > c_v$, the constant volume line is steeper than the constant pressure line (see Fig. 12.3). Previously we have seen that the Carnot cycle can be formed with two reversible adiabatic processes and two reversible isothermal processes. Since, by definition, a reversible adiabatic process is isentropic, then the Carnot cycle on a *T–s* diagram is simply a rectangle, as shown in Fig. 12.4. The heat transfers in and out are marked on this diagram, and from the fact the the reversible heat transfer is given by the area under the curve it is evident that:

$$Q_{in} = T_{high} \Delta s \qquad (12.9)$$

and:

$$Q_{out} = T_{low} \Delta s \qquad (12.10)$$

Here the thermal efficiency, η_{th} of the Carnot cycle is:

$$\eta_{th} = 1 - \frac{Q_{out}}{Q_{in}} = 1 - \frac{T_{low}}{T_{high}} \qquad (12.11)$$

which is the result obtained previously in Chapter 9.

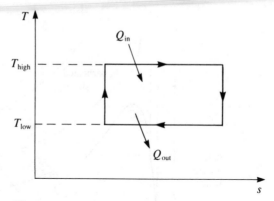

Fig. 12.4   A Carnot cycle on a T–s diagram

12.3   Pressure–volume and temperature–entropy diagrams for vapour–liquid systems

For vapour–liquid systems like steam and water the p–v and the T–s diagrams have to be drawn from tabulated data. However, it is known that superheated vapour at high temperature and low pressure behaves like an ideal gas. Therefore there should, in some regions, be some similarity between the diagrams for a vapour–liquid system and an ideal gas. Figure 12.5 shows the pressure–volume diagram for steam and water, and Fig. 12.6 shows the temperature–entropy diagram. In each figure the different regions for subcooled water, two-phase steam–water mixture, and superheated steam are marked, as is the critical point, C. In the two-phase region as the saturated water evaporates to give saturated steam, the specific volume and the specific entropy both increase, but the pressure and the temperature remain constant. This explains the horizontal lines in both figures in the two-phase regions. It can also be seen that the behaviour of highly superheated steam is beginning to look like the behaviour of an ideal gas:

1. In Fig. 12.5 at high temperatures the isothermal lines are beginning to look like rectangular hyperbolae, especially at low pressures.

2. In Fig. 12.6 at high temperatures in the superheated steam region the isobar lines curve upwards in a similar way to that for ideal gases (see Fig. 12.3).

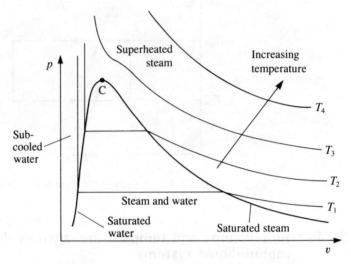

Fig. 12.5 p–v diagram for steam and water

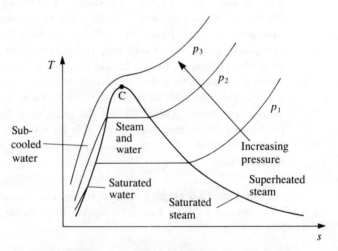

Fig. 12.6 T–s diagram for steam and water

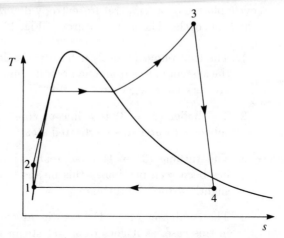

Fig. 12.7   A Rankine cycle on a *T–s* diagram

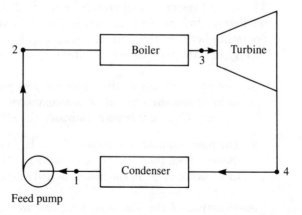

Fig. 12.8   Rankine cycle: block diagram of components

The other main feature of Fig. 12.6 is that in the subcooled region the isobars are extremely close together, indicating that the properties of liquid water are almost independent of pressure. In this region properties such as specific volume, internal energy, enthalpy, and entropy are functions of two other variables of state such as pressure and temperature. However, in practice these properties are strong functions of temperature, but very weak functions of pressure.

The Rankine cycle previously considered in Chapter 7 has already been plotted on a pressure–volume diagram. Figure 12.7 shows the cycle plotted on a temperature–entropy diagram, and Fig. 12.8 shows the basic cycle. The main features of Fig. 12.7 are as follows.

1. The feed pump $(1 \rightarrow 2)$ is hardly visible. As the pressure rises the system moves from one pressure line to another but the lines are very close together.

2. The boiler $(2 \rightarrow 3)$ is a line of constant pressure going from subcooled water to superheated steam.

3. The turbine $(3 \rightarrow 4)$ is adiabatic, but also irreversible. As has been seen previously this means that the entropy increases slightly across the turbine.

4. The condenser $(4 \rightarrow 1)$ operates at constant pressure, and so in this case, as it goes from wet steam to saturated water, is a horizontal constant temperature line.

The advantage of a temperature–entropy diagram is that it is immediately obvious where a cycle is less efficient than a Carnot cycle operating between the same maximum and minimum temperatures. In Fig. 12.7 the reasons are as follows.

1. The irreversibility of the turbine moves point 4 to the right and so increases the amount of heat extracted in the condenser. Increasing Q_{out} in this way decreases the efficiency of the cycle.

2. The heat is added to the cycle over a large temperature range, all the way from the lowest temperature to the highest temperature. In a Carnot cycle all the heat is added at the highest temperature.

Consideration of the shape of temperature–entropy diagram often makes it easier to find the efficiency of a cycle.

Example 12.1   An ideal gas with a constant specific heat c_p, is heated reversibly at constant pressure until its absolute temperature doubles. Then it is expanded reversibly and adiabatically until the temperature returns to the original value. Finally it is compressed isothermally and reversibly until the original state is reached. Find the thermal efficiency of this cycle.

Solution   The cycle is made up of three reversible processes: a isobaric process, an isentropic process, and an isothermal process. The cycle is sketched in Fig. 12.9. The two temperatures are T and $2T$. The heat transfers in each process can be calculated as follows.

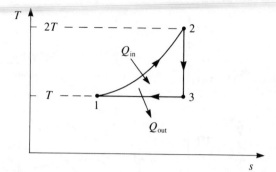

Fig. 12.9 *T–s* diagram for three process cycle: example 12.1

1. Process $1 \to 2$. This is a constant pressure heating process, and so the heat added must be $c_p \Delta T$, hence:

$$Q_{in} = c_p(2T - T) = c_p T \qquad (12.12)$$

2. Process $2 \to 3$. This is an isentropic process, and so the heat transfer is zero.

3. Process $3 \to 1$. This is an reversible isothermal process, and thus the entropy change is the same as in the constant pressure heating. Using one of the equations for entropy change for an ideal gas:

$$\Delta s = c_p \ln \frac{T_f}{T_i} - R \ln \frac{p_f}{p_i} = c_p \ln 2 \qquad (12.13)$$

and so:

$$Q_{out} = T \Delta s = c_p T \ln 2 \qquad (12.14)$$

The thermal efficiency is then easily calculated using eqns 12.12 and 12.14:

$$\eta_{th} = 1 - \frac{Q_{out}}{Q_{in}} = 1 - \ln 2 = 0.307 \qquad (12.15)$$

The efficiency of a Carnot cycle operating between T and $2T$ is $\frac{1}{2}$. The reason why this cycle has a lower efficiency is that all the heat is not added at the top temperature.

12.4 The Mollier chart

The temperature–entropy diagram is useful for seeing the cycle and its characteristics, but for a vapour–liquid system is does not help at all

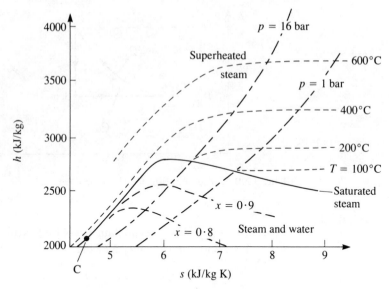

Fig. 12.10 Mollier diagram for steam–water

to perform any of the calculations. The calculations can be performed by the use of steam tables, or equivalent for other substance, but this method is often rather tedious. The property which is often required is enthalpy. This is because, from the steady-flow energy equation, changes in enthalpy are related to work and heat transfers. A diagram is therefore required which gives information about enthalpy. The most commonly used type of diagram is known as the Mollier diagram and is a plot of enthalpy against entropy[1].

The Mollier diagram for steam–water is shown in Fig. 12.10. In the form most commonly used, shown here, the diagram concentrates on the superheated steam region and the wet steam region. It gives no information about saturated water—the critical point will be seen in the bottom left-hand corner of the chart. In the superheated steam region there are two sets of intersecting lines. One set are constant temperature lines, and the other are constant pressure lines. Given superheated steam of known pressure and temperature the relevant point can be identified on the diagram. When the superheated steam reaches a high temperature, the steam behaves ideally.

[1]Note that for an ideal gas with a constant c_p enthalpy is proportional to temperature, and so a Mollier diagram would look just like a temperature–entropy diagram.

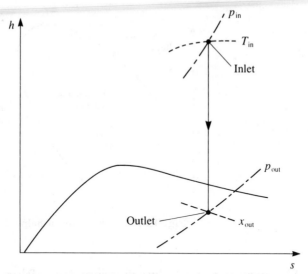

Fig. 12.11   The use of the Mollier diagram for turbine calculations

This can be seen because here the constant temperature lines tend to run horizontally. When this occurs the enthalpy depends only on the temperature—one of the properties of an ideal gas.

In the wet steam region there are again two sets of intersecting lines, but this time they are lines of constant pressure (which are continuous with those in the superheated steam region) and lines of constant dryness, x. Constant temperature lines are not given in this region because the pressure is determined by the temperature and so the constant pressure lines are also a constant temperature lines. The value of the temperature along a particular constant pressure line can be seen by looking at the temperature along the saturated steam line as indicated in Fig. 12.10. The particular use of a Mollier diagram is for turbine calculations (see Fig. 12.11). The superheated steam conditions at the turbine inlet are easily located from the temperature and the pressure, and the enthalpy can be read off. If the turbine is isentropic, then a line a drawn vertically downwards until the required turbine outlet pressure is reached. From the resulting point the enthalpy and the dryness of the steam can be read off.

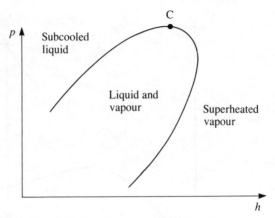

Fig. 12.12   Pressure–enthalpy diagram

12.5   The pressure–enthalpy diagram for refrigeration systems

As will be seen later, refrigeration systems are cycles which operate at two distinct pressures. Fluid flows round the cycle being compressed from the lower pressure to the higher. Then later in the cycle the pressure is allowed to fall suddenly back to the lower pressure. The advantage of the pressure–enthalpy diagram (see Fig. 12.12) is that these two pressure regions are represented as two horizontal lines.

The lines shown on pressure–enthalpy diagrams are illustrated in Fig. 12.12. This diagram covers the whole of the fluid range from subcooled liquid to superheated vapour. It contains a saturated liquid line and a saturated vapour line, with the critical point at the meeting of these lines.

12.6   Problems

12.1 An ideal gas (at state 1) with a constant specific heat c_v, is heated reversibly at constant volume until its absolute temperature doubles (state 2). Then it is expanded reversibly and adiabatically until the temperature returns to the original value (state 3). Finally it is compressed isothermally and reversibly until the original state (state 1) is reached. Find the thermal efficiency of this cycle. Sketch the T–s diagram for this cycle. [This cycle is very similar to that in example 12.1.]

12.2 In question 12.1 find:

(a) an equation for the net work output W in terms of c_v and T_1;

(b) the value of the ratio v_3/v_2. (Note that $v_1 = v_2$.)

(c) the value of the maximum volume change, Δv_{max} in terms of v_1;

(d) the value of the ratio $W/\Delta v_{max}$ in terms of the pressure of the original state, p_1. This ratio is the mean effective pressure of the cycle—see Chapter 17.

Draw the cycle from question 12.1 on a p–v diagram, and interpret the mean effective pressure graphically.

12.3 The cycles in question 12.1 and example 12.1 have the same thermal efficiency but different values of the mean effective pressure. How can the value of the mean effective pressure be used to decide which cycle is preferential?

12.4 An ideal gas with a constant specific heat is heated reversibly at constant pressure until its absolute temperature doubles. Then it is cooled reversibly at constant volume until the temperature returns to the original value. Finally it is compressed isothermally and reversibly until the original state is reached. Find the thermal efficiency of this cycle. Sketch the T–s diagram for this cycle.

12.5 For the cycle in question 12.4 find the mean effective pressure in terms of p_1.

12.6 The Carnot cycle consists of two isothermals and two isentropics. The Stirling cycle consists of two isothermals and two constant volume lines. The Ericsson cycle consists of two isothermals and two constant pressure lines. All these processes are reversible, and the cycles are run as heat engines: to convert heat into work. Sketch the three cycles, for ideal gases, on p–v and T–s diagrams. Each cycle has a high-temperature isothermal at T_h and a low temperature isothermal at T_l. Prove that all three cycles have the same efficiency if, for the Stirling and the Ericsson cycles, heat liberated in the cycle can be re-used in the cycle. For the case where $T_h = 2T_l$, and where the minimum volume in the cycle is 10% of the maximum volume, calculate the mean effective pressure of each cycle in terms of p, the pressure at the beginning of the isothermal compression.

12.7 From the results of question 12.6 explain why the Stirling cycle has received more attention for practical development than the Carnot or Ericsson cycles.

12.8 The inlet steam condition to a turbine is 100 bar and 500°C. The outlet condition is 0.04 bar and $x = 0.85$. Assuming the steam follows a straight line on an h–s diagram, find the pressure at which dry saturated steam can be bled off part way along the turbine.

12.9 An isentropic turbine expands steam from 500°C to an outlet pressure of 0.1 bar or $x = 0.9$, whichever occurs first. Find the inlet pressure which maximizes the work output per kilogram of steam, and find this maximum work output.

13

Turbines and compressors

13.1 Key points of this chapter

- For practical reasons turbines are usually adiabatic. Maximum power is produced when the turbine is also reversible. (Section 13.2)

- For an ideal gas, the outlet conditions for a reversible, adiabatic turbine can be calculated. A turbine isentropic efficiency to allow for irreversibilities can be defined. (Section 13.3)

- For steam turbines it is necessary to use steam tables or a Mollier chart to find the outlet conditions and therefore the power produced. (Section 13.4)

- Steam turbines often have wet steam at the exit. The water drops can sometimes damage the turbine. The turbine or the whole cycle can be modified to prevent damage occurring. (Section 13.5)

- Compressors for gases can be treated in a similar way to turbines. For an ideal gas the equations are simple, and a compressor isentropic efficiency can be defined. (Section 13.6)

- Compressors for saturated liquids are used in refrigeration cycles. The use of refrigeration tables is explained. (Section 13.6)

- An alternative criterion of compressor efficiency, the isothermal efficiency is defined. Compressors use less power if attempts are made to cool the gas and keep it at constant temperature. (Section 13.7)

13.2 Turbines: general principles

A turbine takes high-pressure, high-temperature fluid and extracts work from it as the pressure falls. Turbines can operate with liquids or with gases and are steady-flow machines.

Turbines operating with liquids are commonly found in hydro-electric power stations. They use water at a high pressure drawn from an elevated reservoir to generate power. The flow in these machines is essentially at constant temperature, and the machines can be analyzed by the techniques of fluid mechanics. Such machines are not considered further here.

Practical turbines operating with the gaseous phase are either steam turbines or gas turbines. The differences, as far as we are concerned, are not great. Gas turbines are often analyzed on the basis that the gas is ideal, whereas steam turbines have to be analyzed with tabulated or graphical steam properties. Often, at the outlet from a steam turbine, the 'steam' is actually a mixture of water droplets and steam.

Whatever type of turbine is considered, the starting point for the analysis is the steady-flow energy equation. Writing the equation per kilogram of fluid flowing, and ignoring the kinetic energy and the potential energy terms:

$$Q - W_s = \Delta h \tag{13.1}$$

where Q is the heat transfer to the system per kg of fluid, W_s is the shaft work done by the system per kg of fluid, and Δh is the change in specific enthalpy.

The aim of a turbine is to produce as much work as possible, that is to maximize W_s. Both steam turbines and gas turbines are usually much hotter than their surroundings and so they tend to lose heat to the surroundings, that is Q is negative. Since $W_s = Q - \Delta h$, a negative value for Q reduces the work produced. In practice the best that can be hoped for is that the turbine will be thermally insulated so that $Q = 0$. Therefore eqn 13.1 is almost always abbreviated to:

$$W_s = -\Delta h \tag{13.2}$$

Turbines are therefore assumed to be adiabatic, and so the entropy change of the surroundings is equal to zero. It could, in theory, still be reversible or irreversible, and so the overall entropy change could be respectively zero or positive. Thus the entropy change of the fluid is also zero or positive.

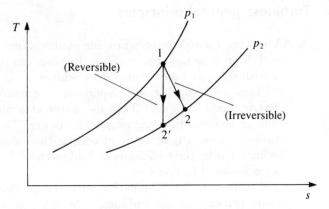

Fig. 13.1 Temperature–entropy diagram for an ideal gas

The situation for an ideal gas (see Fig. 13.1) is most easily seen using a temperature–entropy diagram.

The gas is falling in pressure from a high pressure (p_1) to a low pressure (p_2), so in Fig. 13.1 we are moving from the higher constant pressure line to the lower one. For an ideal gas the enthalpy change is proportional to the temperature change. Since we want to maximize the work output, this means, from eqn 13.2, maximizing the enthalpy change and so maximizing the temperature change. The turbine is adiabatic, and so in Fig. 13.1 a reversible change is an isentropic change, and an irreversible change involves an increase in entropy of the working fluid. These two possibilities are shown in Fig. 13.1. It is now quite clear that to maximize the temperature change and so maximize the work output, the turbine should by isentropic.

This is an important conclusion. Turbines are assumed in practice to be adiabatic, and the work output is maximized if the turbine is isentropic.

13.3 Gas turbines

A turbine operating with an ideal gas is not difficult to analyze. Figure 13.2 shows a temperature–entropy diagram similar to Fig. 13.1. Point $2'$ is vertically below point 1, so the process $1 \rightarrow 2'$ is an isentropic process. Because the gas is ideal:

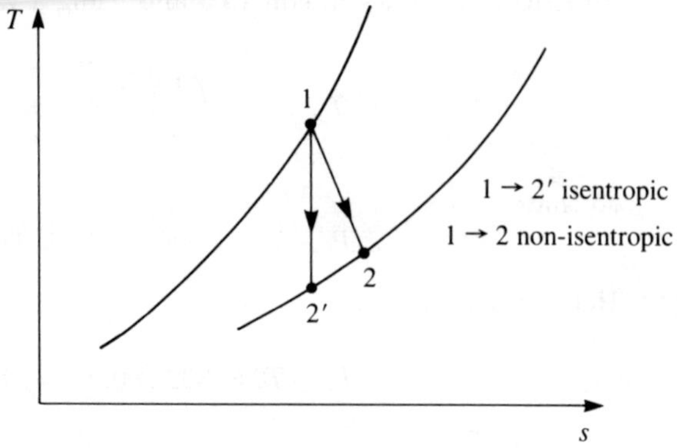

Fig. 13.2 Isentropic and non-isentropic gas turbines

$$T_{2'} = T_1 \left(\frac{p_2}{p_1} \right)^{\frac{\gamma-1}{\gamma}} \tag{13.3}$$

and the work produced by this isentropic turbine $W_{s,\text{isentropic}}$ is:

$$W_{s,\text{isentropic}} = -\Delta h = c_p(T_1 - T_{2'}) \tag{13.4}$$

However, the real process is not isentropic, and the end point of the process is not 2′ but 2 in Fig. 13.2. A sensible definition of turbine efficiency η_T is then:

$$\eta_T = \frac{W_{s,\text{actual}}}{W_{s,\text{isentropic}}} \tag{13.5}$$

This implies, considering eqn 13.4, that:

$$\eta_T = \frac{T_1 - T_2}{T_1 - T_{2'}} \tag{13.6}$$

Because this efficiency is defined in terms of the isentropic work output it is often known as the **turbine isentropic efficiency**. In a typical calculation the pressures p_1 and p_2 are known, as are the turbine inlet temperature T_1 and the isentropic efficiency η_T. It is then straightforward to calculate the actual work output from the turbine.

Example 13.1

If $p_1 = 5$ bar, $p_2 = 1$ bar, $T_1 = 900$ K, and $\eta_T = 90\%$, calculate the values of $T_{2'}$, T_2, and $W_{s,\text{actual}}$. Assume the gas has the properties of air.

Solution   Substituting values in eqn 13.3 and using a value of 1.4 for γ:

$$T_{2'} = 900 \left(\frac{1}{5}\right)^{\frac{1.4-1}{1.4}} = 568 \text{ K}$$

so that:

$$T_1 - T_{2'} = 900 - 568 = 332 \text{ K}$$

Hence from eqn 13.6:

$$T_1 - T_2 = 332 \times 0.90 = 299 \text{ K}$$

and so $T_2 = 601$ K. Using a value of 1.005 kJ/kg K for c_p, the value of $W_{s,\text{actual}}$ can then be calculated from:

$$W_{s,\text{actual}} = c_p(T_1 - T_2) = 1.005 \times (900 - 601) = 300 \text{ kJ/kg}$$

The value of 90% for the isentropic efficiency is a typical value for a modern gas turbine.

13.4   Steam turbines

For steam turbines similar principles hold. The turbine which produces the maximum power output is the isentropic turbine, and an isentropic efficiency η_T can be defined by eqn 13.5. Of course the steam does not behave ideally, and indeed many steam turbines have a mixture of steam and water droplets at the exit (see Fig. 13.3). Once again the isentropic turbine is represented by the process $1 \rightarrow 2'$, and the actual non-isentropic turbine by $1 \rightarrow 2$. Figure 13.3 shows the case where the turbine exhaust is wet steam. In eqn 13.5 the following substitutions can be made because the turbine is still assumed to be adiabatic:

$$W_{s,\text{actual}} = h_1 - h_2 \tag{13.7}$$

and:

$$W_{s,\text{isentropic}} = h_1 - h_{2'} \tag{13.8}$$

to give:

$$\eta_T = \frac{h_1 - h_2}{h_1 - h_{2'}} \tag{13.9}$$

Equation 13.9 is expressed in terms of a ratio of enthalpy differences rather than a ratio of temperature differences because of course enthalpy can only be replaced by temperature for an ideal gas.

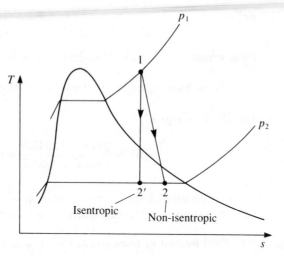

Fig. 13.3   Isentropic and non-isentropic steam turbines

The actual calculation procedure is now best explained by an example.

Example 13.2   The turbine in the Rankine cycle example in Chapter 7 had the following inlet and outlet conditions:

inlet   1 $T = 500°C$ $p = 100$ bar
outlet   2   dryness, $x = 0.8$ $p = 0.0233$ bar

The turbine is adiabatic: calculate the turbine isentropic efficiency.

Solution   From the enthalpy tables in Chapter 7:

$$h_1 = 3374.6 \text{ kJ/kg}$$

and:

$$h_2 = h_{f2} + x_2 h_{fg2} = 83.9 + 0.8 \times 2454.3 = 2047.3 \text{ kJ/kg}$$

It is now necessary to calculate $h_{2'}$ remembering that 2′ has the same entropy as 1. From the entropy tables in Chapter 10:

$$s_1 = 6.5994 \text{ kJ/kg K}$$

It is also known that:

$$s_{2'} = s_1 = s_{f2} + x_{2'} s_{fg2}$$

or:
$$6.5994 = 0.2963 + x_{2'} \times 8.3721$$

from which $x_{2'} = 0.753$. This dryness can be used to find $h_{2'}$:

$$h_{2'} = h_{f2} + x_{2'} h_{fg2} = 83.9 + 0.753 \times 2454.3 = 1931.7 \text{ kJ/kg}$$

Finally from eqn 13.9:

$$\eta_T = \frac{3374.6 - 2047.3}{3374.6 - 1931.7} = 0.92$$

A more usual form of the calculation is given p_1, T_1, p_2, and η_T to find x_2 and h_2. The procedure is then as follows.

1. Find h_1 and s_1 from steam tables or a Mollier chart.

2. Put $s_{2'} = s_1$ and so find $x_{2'}$ either directly from a Mollier chart or from:
$$s_{2'} = s_{f2} + x_{2'} s_{fg2} \qquad (13.10)$$

3. Find $h_{2'}$ either directly from a Mollier chart or from:
$$h_{2'} = h_{f2} + x_{2'} h_{fg2} \qquad (13.11)$$

4. Find h_2 from eqn 13.9.

5. Find x_2 from:
$$h_2 = h_{f2} + x_2 h_{fg2}$$

It should be noted that the use of the Mollier chart simplifies and speeds up the calculation considerably.

13.5   The effect of liquid droplets on steam turbines

When the steam at the outlet of a turbine becomes wet, the liquid present is in the form of water droplets. If the steam is not very wet, the amount of water is small and the droplets are not too troublesome to the operation of the turbine. This is because the liquid drops are formed by condensation from the steam to form a kind of fog. Like atmospheric fog, this fog contains extremely small drops and the drops move with almost the same velocity as the surrounding steam.

However as the dryness of the steam decreases, the concentration of these water droplets increases. The turbine blades move rapidly

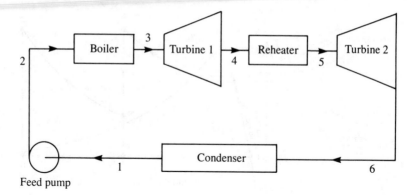

Fig. 13.4   Rankine cycle with reheat

through the steam and tend to collect the water droplets. This is because the water droplets do not move with quite the same velocity as the steam, and so get 'scooped up' by the blade. Once on the blade the water forms a film and runs to the back of the blade. Here the water is re-entrained into the steam. But this droplet formation method is very different from the original one. The droplets are a completely different size: they are much larger and now no longer follow the steam flow.

When these large drops impact with the turbine blades they can do much damage and certainly impair the efficiency of the turbine. There are two remedies available for this problem.

1. Remove the water from the turbine. This can be done by sucking water from the turbine casing or from holes in the turbine blades. If this is not done it is often considered unwise to work with drynesses of less than around 85% to 90%.

2. Designing the cycle so that the steam at the turbine exit is not very wet. This is not as difficult as it sounds. Consider the cycle shown in block diagram form in Fig. 13.4. It is a conventional Rankine cycle except that there are two turbines with an additional heating process (the 'reheat' stage) between them. The temperature–entropy diagram for this cycle is shown in Fig. 13.5. Note how the effect of adding the reheat to the cycle has moved the point representing the end of the low pressure turbine to the right (from 4' to 6), and has therefore made the steam drier. Later we will see that the reheat has improved the cycle efficiency

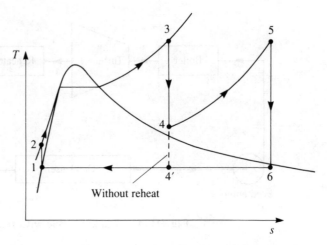

Fig. 13.5   *T–s* diagram for a Rankine cycle with reheat

slightly. Just as valuable as this improvement in efficiency is its contribution to making the steam drier.

13.6   Compressors and isentropic efficiency

The compression of an ideal gas is often considered in an analogous way to that for a turbine (see Section 13.3). In other words the compressor is assumed to be adiabatic, and the goal is isentropic operation. Hence in Fig. 13.6 the process $1 \rightarrow 2'$ is an isentropic process, and the process $1 \rightarrow 2$ is an irreversible, adiabatic process. The **compressor isentropic efficiency** η_C is then defined to be:

$$\eta_C = \frac{W_{s,\text{isentropic}}}{W_{s,\text{actual}}} \qquad (13.12)$$

or that:

$$\eta_C = \frac{T_{2'} - T_1}{T_2 - T_1} \qquad (13.13)$$

It should be noted that there is a subtle difference between these equations and the ones for turbine efficiency—eqns 13.5 and 13.6. The difference is that here the isentropic terms are in the numerator; previously they were in the denominator. This is because the isentropic compressor uses the minimum power, whilst the isentropic turbine produces the maximum power. The definitions of efficiency are arranged to be always less than 1.

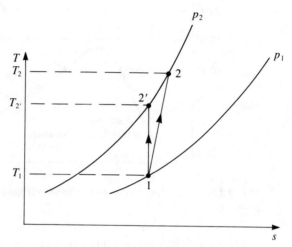

Fig. 13.6   Compression of an ideal gas

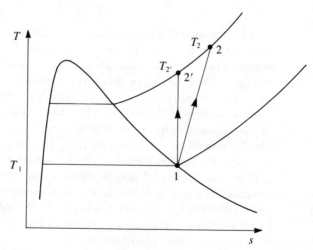

Fig. 13.7   Compression of a saturated vapour: example 13.3

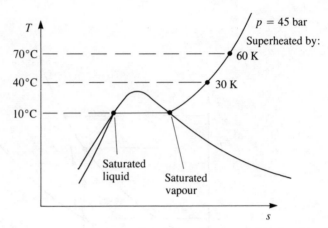

Fig. 13.8   Constant pressure line from refrigerant table showing the four points given

Table 13.1   Refrigerant table, extract for carbon dioxide

p (bar)	Saturated T_{sat} (°C)	h_f (kJ/kg)	h_g (kJ/kg K)	s_f (kJ/kg)	s_g (kJ/kg K)	Superheated by 30 K h (kJ/kg)	Superheated by 30 K s (kJ/kg K)	Superheated by 60 K h (kJ/kg)	Superheated by 60 K s (kJ/kg K)
12.0	-35	9.7	322.2	0.039	1.352	356.9	1.485	385.6	1.588
45.0	10	109.1	307.2	0.407	1.107	364.7	1.302	401.4	1.414

Compression of a liquid–vapour mixture is not possible in practice. However, compression of a saturated vapour does commonly occur in refrigeration systems. This process can be handled by assuming that the gas is ideal, but often this is not a good assumption even though the compression takes the vapour away from the saturation line and into the superheated region (see Fig. 13.7). The use of tabulated data in this region requires some explanation. The general form of the tables for refrigerants is shown in Table 13.1, which is an extract from the table for carbon dioxide. The important point about this table is that each horizontal line all refers to the same pressure and gives information about four points at this pressure: saturated liquid, saturated vapour, vapour superheated by 30 K, and vapour superheated by 60 K. These points are illustrated in Fig. 13.8.

The use of this type of table is best illustrated by an example.

Example 13.3   Saturated carbon dioxide vapour at 12 bar is compressed to 45 bar. If the compressor isentropic efficiency is 85%, calculate the exit temperature.

Solution Table 13.1 gives the required properties, and Fig. 13.7 shows the compression on a T–s diagram. From Table 13.1 the inlet conditions are:

$$h_1 = 322.2 \text{ kJ/kg} \quad \text{and} \quad s_1 = 1.352 \text{ kJ/kg K}$$

First find the point $2'$ where the pressure is 45 bar and $s_{2'} = s_1$. The entropy at a superheat of 30 K is too low (1.302 kJ/kg K) and at a superheat of 60 K it is too high (1.414 kJ/kg K). Therefore interpolate linearly between these points to find the superheat at $2'$:

$$\text{superheat} = 30 + 30\frac{1.352 - 1.302}{1.414 - 1.302} = 43.39 \text{ K}$$

and then from this superheat find the enthalpy at $2'$:

$$h_{2'} = 364.7 + \frac{13.39}{30}(401.4 - 364.7) = 381.1 \text{ kJ/kg}$$

The relevant equation for η_C is:

$$\eta_C = \frac{h_{2'} - h_1}{h_2 - h_1} \tag{13.14}$$

In this equation, only h_2 is unknown. Substituting the other values: $\eta_C = 0.85$, $h_1 = 322.2$ kJ/kg, and $h_{2'} = 381.1$ kJ/kg, the value of h_2 can be found to be 391.5 kJ/kg. Re-interpolating to find the corresponding temperature, the superheat can be shown to be 51.9 K, and so the outlet temperature is 61.9°C.

It is interesting to see the result if the gas is assumed to be ideal, and the known value of γ used. For carbon dioxide the experimental value of γ is 1.29. Using this value the result for $T_{2'}$ is 47.3°C compared with the result from tables of 53.4°C. The vapour in this case is behaving almost like an ideal gas.

13.7 Compressors: another definition of efficiency

For compressors in a refrigeration system which operate below ambient temperature, the best that can be hoped for is that the compressor will be adiabatic, since heat transfer to the compressor increases the compression work. Similar reasoning to that used in Section 13.3 leads to the conclusion that the compressor should be reversible as well as adiabatic, therefore the isentropic compressor is the ideal. In

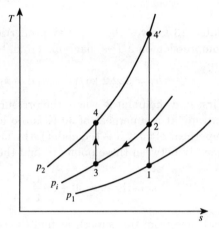

Fig. 13.9 Two-stage compression process: example 13.4

this case the use of the isentropic efficiency as defined above is quite justified. The isentropic compression is the best that can be achieved; however it cannot actually be achieved. It is not possible to perform the process reversibly and adiabatically. Adiabatic operation is not too demanding, but reversible operation certainly cannot be achieved.

In a compressor operating above atmospheric pressure there is certainly no reason in theory why the heat transfer cannot be negative (heat removed from the compressor) and therefore the work of compression reduced. It is easiest to see this by means of a simple example of a two-stage compression process.

Example 13.4

An ideal gas is compressed isentropically from p_1 to p_2 in two stages. The first compressor increases the pressure to an intermediate pressure p_i. After this initial compression the gas is cooled to its original temperature. The final compressor raises the pressure to p_2. Find an equation for the optimum intermediate pressure. If the gas is air, the initial temperature is 300 K, and the pressure is raised from 1 bar to 10 bar; find the work saved by the cooling process.

Solution

Figure 13.9 shows the compressor system on a T–s diagram. The process $1 \rightarrow 4'$ represents a single stage-compression. The processes $1 \rightarrow 2$ and $3 \rightarrow 4$ represent the two-stage compression process. The compression work is proportional to the temperature rise in each compressor, or to the length of the vertical lines in Fig. 13.9. It is evident that the total work of the two-stage process is less than that in the single-stage process because the lines of constant pressure on a T–s diagram diverge as the temperature increases. For an isentropic process the final temperature in a compression is:

$$T_2 = T_1 \left(\frac{p_i}{p_1}\right)^{\frac{\gamma-1}{\gamma}} \tag{13.15}$$

and the temperature rise $T_2 - T_1$ is given by:

$$T_2 - T_1 = T_1 \left[\left(\frac{p_i}{p_1}\right)^{\frac{\gamma-1}{\gamma}} - 1\right] \tag{13.16}$$

The compression work is proportional to this temperature difference. For the two-stage compression process there are two terms similar to the right hand side of eqn 13.16:

$$(T_2 - T_1) + (T_4 - T_3) = T_1 \left[\left(\frac{p_i}{p_1}\right)^{\frac{\gamma-1}{\gamma}} - 1 + \left(\frac{p_2}{p_i}\right)^{\frac{\gamma-1}{\gamma}} - 1\right] \tag{13.17}$$

remembering that since the gas is cooled back to its original temperature after the first compression, $T_3 = T_1$. Equation 13.17 can be differentiated with respect to p_i so that the minimum work is found. This is not as difficult as it looks if the equation is actually differentiated with respect to $p_i^{(\gamma-1)/\gamma}$. The minimum occurs when:

$$p_i = \sqrt{p_1 p_2} \tag{13.18}$$

For the single-stage process, from eqn 13.16:

$$\Delta T = 300 \left[\left(\frac{10}{1}\right)^{\frac{1.4-1}{1.4}} - 1\right] = 279 \text{ K}$$

and the work is $c_p \Delta T = 1.005 \times 279 = 280$ kJ/kg. For the two-stage process, from eqn 13.17 the total temperature rise in the two compressors is given by:

$$\Delta T = 300 \left[2 \left(\frac{\sqrt{10}}{1}\right)^{\frac{1.4-1}{1.4}} - 2\right] = 234 \text{ K}$$

and the work is $c_p \Delta T = 1.005 \times 234 = 235$ kJ/kg.

By using the two-stage process with cooling between the stages (**intercooling**) the compression work has been reduced by 16%.

The obvious logical conclusion is that the work can be further reduced by compressing the gas reversibly and isothermally. In this case the work can be calculated from:

$$W_s = - \int v dp \qquad (13.19)$$

remembering that, since this is an isothermal process with an ideal gas, $pv = RT$. The result of the integration is that:

$$W_s = -RT \ln \left(\frac{p_f}{p_i} \right) \qquad (13.20)$$

The 'minus' sign in this equation indicates that the compression work has to be done on the system. Substituting values for the above example for air, the compression work is $0.287 \times 300 \times \ln 10 = 198$ kJ/kg. This is the isothermal work, $W_{isothermal}$. An **isothermal efficiency** can now be defined as:

$$\eta_{isothermal} = \frac{W_{isothermal}}{W_{actual}} \qquad (13.21)$$

On this basis the isothermal efficiency of the two-stage compressor above is:

$$\eta_{isothermal} = \frac{198}{235} = 0.84$$

The problem now is that there are two alternative definitions of efficiency for a compressor.

1. The isentropic efficiency which relates the work to the work required in an isentropic compressor.

2. The isothermal efficiency which relates the work to the work required in an reversible isothermal compressor.

The isothermal work is always lower; indeed it is not difficult to show that the ratio of the reversible isothermal work to the isentropic work, which can be interpreted as the isothermal efficiency of an isentropic compressor, is:

$$\eta_{isothermal} = \frac{\gamma - 1}{\gamma} \frac{\ln p_r}{p_r^{\frac{\gamma-1}{\gamma}} - 1} \qquad (13.22)$$

where p_r is the pressure ratio for the compressor, that is the ratio of the outlet pressure to the inlet pressure. For $p_r = 10$ and $\gamma = 1.4$, this has a value of 0.707. It is also quite possible for a compressor to have an isentropic efficiency greater than 1. For the values quoted above, the isentropic efficiency of a reversible isothermal compressor is $1/0.707 = 1.415$.

It is clear that the appropriate definition of efficiency should be used. The isentropic efficiency should be used where the compressor is adiabatic, but irreversibilities mean that it is not isentropic. The isothermal efficiency should be used where the compressor is approximately reversible, but incomplete heat transfer means that it is not isothermal. In practice of course real compressors are neither adiabatic nor reversible.

13.8 Problems

13.1 An turbine with isentropic efficiency of 90% expands air from 10 bar and 500°C to 1 bar. Calculate the final temperature of the air, and the work output per kilogram of air flowing.

13.2 Repeat question 13.1 with steam. Assume the superheated steam behaves as an ideal gas with $\gamma = 1.3$.

13.3 Repeat question 13.2 using steam tables rather than ideal gas theory. Compare the answers with those obtained previously. Was the assumption of ideal gas behaviour a good one?

13.4 If the steam is now expanded from 10 bar and 500°C to 0.04 bar with isentropic efficiency of 90%, in what respects does the ideal gas assumption become invalid. Again compare the ideal gas and steam table results for the exit temperature and the work output.

13.5 Steam is expanded isentropically in a turbine from 100 bar and 500°C to 0.04 bar. Find the outlet condition of the steam and the work produced per kilogram of steam flowing. Explain why this turbine would not be satisfactory.

13.6 The steam expansion in question 13.5 is performed until the steam dryness falls to 90%. At this point 90% of the water is extracted from the steam, and the expansion continued. Again the expansion is stopped when the dryness falls to 90%, and again 90% of the water is extracted. The steam is then allowed to expand to 0.04 bar. Calculate the final condition of the steam and find the total work output per kilogram of steam input to the turbine. Comment on the result.

13.7 An ideal gas is compressed isentropically from p_1 to p_2 in a three-stage compressor. Between the stages the gas, which has the properties of air, is cooled to its original temperature. Prove (for example using Lagrange's method of multipliers) that the work is minimized if the stage pressure ratios are equal, and the the stage pressure ratio is the cube root of p_2/p_1. If p_1 is 1 bar, p_2 is 10 bar and the inlet temperature of the air is 300 K, find the isothermal efficiency of the compressor.

Compare this isothermal efficiency with the isothermal efficiency of a single-stage isentropic compressor.

13.8 Saturated ammonia vapour at 1.9 bar is compressed to 10 bar in a compressor with an isentropic efficiency of 90%. Find the outlet temperature and the work of compression per kilogram of ammonia.

14

Steady-flow power cycles: the Rankine cycle

14.1 Key points of this chapter

- The simplest form of saturated Rankine cycle has a high work ratio but a disappointingly low thermal efficiency. (Section 14.2)

- The thermal efficiency can be improved, in theory, by making the cycle more like a Carnot cycle, and by increasing the maximum temperature of the cycle. (Section 14.3)

- The main way the cycle can be made more like a Carnot cycle is to add more of the heat near the maximum temperature. (Section 14.3)

- The maximum temperature can be increased by increasing the pressure, and by superheating the steam. (Section 14.4)

- More of the heat can be added near the maximum temperature by re-heating the steam and having two turbines, and by bleeding off part of the steam to heat the feed water. (Section 14.4)

- Re-heating the steam also has the advantage of making the steam drier at the turbine exit. (Section 14.4)

14.2 The basic Rankine cycle

The Rankine cycle has already been mentioned in Chapter 7. It is the most common thermodynamic cycle for large-scale power production from heat, and has been used in one form or other since the earliest days of steam power. The simplest cycle is shown as a block diagram

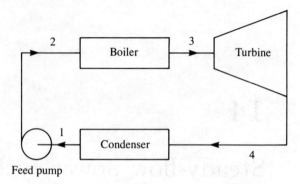

Fig. 14.1   Block diagram of a simple Rankine cycle

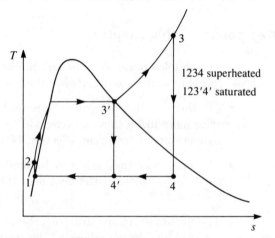

Fig. 14.2   Rankine cycle (saturated and superheated): temperature–entropy diagram

in Fig. 14.1. The cycle is also shown as a temperature–entropy diagram in Fig. 14.2. As will be seen from this Figure, the turbine is, for the moment, assumed to be isentropic. This type of cycle may be analyzed by the methods from Chapter 7. The results of an analysis are as follows.

1. The thermal efficiency is disappointingly low, perhaps only 30%.

2. The work ratio is high, almost equal to 1. This is because the work input to the cycle, in the boiler feed pump, is very low. The feed pump work is low because the feed pump is compressing

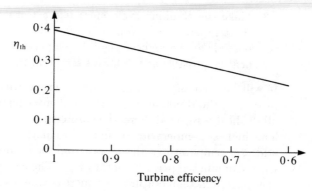

Fig. 14.3 Variation of Rankine cycle efficiency with turbine efficiency: saturated cycle, 100 bar

liquid water, and only a small amount of work produces a very large increase in the pressure.

The significance of the high work ratio is that the overall cycle efficiency is not very sensitive to the efficiency of the feed pump and the turbine. The feed pump work is so small that the efficiency of the feed pump can be neglected. The variation of the overall efficiency with turbine isentropic efficiency is shown in Fig. 14.3.

14.3 Improvements to the Rankine cycle: theory

The main ways of improving the Rankine cycle can be seen by comparing it with the Carnot cycle: the most efficient cycle operating between two temperature T_{high} and T_{low}. The Carnot thermal efficiency η_{Carnot} is:

$$\eta_{Carnot} = 1 - \frac{T_{low}}{T_{high}} \tag{14.1}$$

Hence the way to improve the efficiency of a Carnot cycle is to increase T_{high} and to decrease T_{low}. Consequently the ways to improve the efficiency of the Rankine cycle are as follows.

1. Increase the maximum temperature of the cycle.

2. Decrease the minimum temperature of the cycle. In fact not much can be done about this as the Rankine cycle already rejects its heat at a low temperature—only just above ambient.

3. Make the Rankine cycle more like a Carnot cycle in shape on the temperature–entropy diagram: hence add as much heat as possible at the highest possible temperature, and reject as much heat as possible at the lowest temperature.

It will be seen that nothing much is wrong with the heat rejection process in the Rankine cycle. The condenser temperature is low, and all the heat is rejected there. The problem is in the heat input process. The highest temperature is not particularly high, and the heat is added at temperatures all the way from the condenser temperature to the maximum temperature of the cycle. The message is clear: increase the maximum temperature, and increase the average temperature at which heat is added.

14.4   Improvements to the Rankine cycle: practical improvements

These improvements can be seen as a logical sequence: increasing the temperature and pressure in a saturated cycle as in Fig. 14.2, superheating the steam, superheating and reheating the steam, and finally heating the feed water before it enters the boiler. Of these four measures, the first two are aimed at increasing the maximum temperature, and the final two are aimed at increasing the average temperature. These steps are described below, together with their effect on the cycle efficiency.

Increasing the pressure and temperature in a saturated cycle. We are dealing with the cycle of Figs. 14.1 and 14.2. The variation of thermal efficiency with steam outlet pressure is shown in Fig. 14.4. As the outlet pressure (and therefore the temperature) increases, the thermal efficiency also increases as expected.

Superheating the steam. The superheated steam cycle is similar to Fig. 14.1, and the temperature–entropy diagram is shown in Fig. 14.2. For a particular steam pressure, here 100 bar, the variation of thermal efficiency with the maximum steam temperature is shown in Fig. 14.5. Again as expected the higher the temperature the higher is the efficiency. However, practical considerations about the strength of the material of the boiler tubes limit the maximum temperature at high pressure to around 600°C.

Reheating as well as superheating the steam. Figure 14.6 shows a superheated cycle with reheat. Here there are two turbines: high-pressure and low-pressure turbines. Between these turbines the steam is fed back to the boiler and reheated, usually back to the original temperature. The temperature–entropy diagram is shown in Fig. 14.7. Taking a boiler pressure of 100 bar and a maximum temperature of

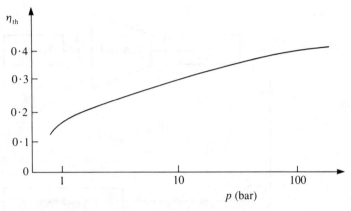

Fig. 14.4   Variation of saturated Rankine cycle efficiency with pressure

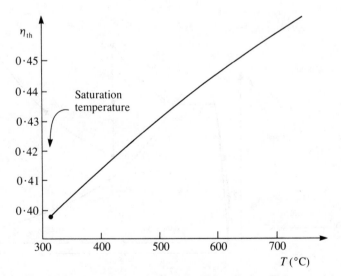

Fig. 14.5   Variation of superheated Rankine cycle efficiency with temperature: pressure = 100 bar

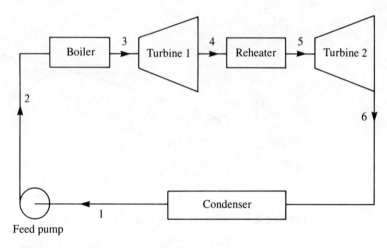

Fig. 14.6 Superheated Rankine cycle with reheat of the steam

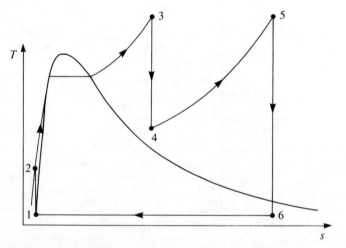

Fig. 14.7 Superheated Rankine cycle with reheat: temperature–entropy diagram

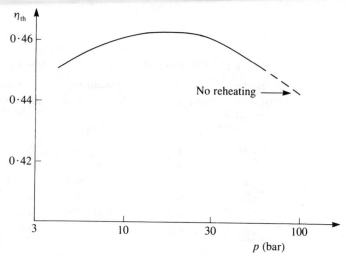

Fig. 14.8 Efficiency of a superheated Rankine cycle with reheat: effect of reheat pressure: boiler pressure = 100 bar, maximum temperature = 600°C

600°C, Fig. 14.8 shows the variation of thermal efficiency as a function of the pressure at the end of the high-pressure turbine. From Fig. 14.8 it can be seen that the use of reheat increases the cycle efficiency. However, the increase in efficiency brought about by the use of reheat is relatively modest. The real benefit of reheat is that is makes the steam at the exit of the low pressure turbine less wet. In Chapter 13 it was emphasized that water droplets at the end of the turbine could damage turbine and decrease efficiency. Figures for the dryness at the turbine exit are given in Table 14.1 for a number of cycles. Cycles having a high efficiency have unacceptably wet steam at the end of the turbine unless re-heating is used.

Regenerative feed-water heating. Re-heating the steam increases the efficiency (albeit by a very small amount) because it increases the average temperature at which heat is added to the cycle. Adding extra heat at the top end of the temperature range produces disappointing results. The alternative is to raise the average temperature by doing something about the low temperature end. A lot of the inefficiency arises because the feed water to the boiler has to be heated from a very low temperature. Regenerative feed-water heating tackles this problem by bleeding off some of the steam part-way through its expansion and using it to heat the water initially. Figure 14.9 shows

Table 14.1   Efficiency and turbine exit dryness for a number of cycles

Boiler pressure (bar)	Boiler exit temperature (°C)	Steam re-heat T & p (°C),(bar)	Cycle efficiency	Dryness at turbine exit
1	saturated	no re-heat	0.178	0.87
100	saturated	no re-heat	0.397	0.65
100	400	no re-heat	0.412	0.72
100	600	no re-heat	0.441	0.81
100	600	600, 50	0.454	0.85
100	600	600, 15	0.461	0.93
100	600	600, 5	0.454	0.99

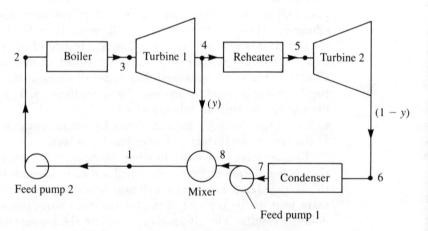

Fig. 14.9   Direct regenerative feed-water heating: block diagram

a block diagram of 'direct' regenerative feed-water heating. 'Direct' here means that the steam is directly mixed with feed water to produce the heating. In the alternative 'indirect' heating method the heat is transferred from the steam to the feed water in a heat exchanger. Indirect heating is thus more complicated, but can avoid the multiple feed pumps shown in Fig. 14.9.

Here the steam is bled off after the high pressure turbine, before reheating. In the cycle shown in Fig. 14.9 this steam is at a temperature of approximately 170°C and a pressure of 5 bar. This steam is directly mixed with water at the same pressure. As the amount of steam fed to the mixer is gradually increased, the outlet temperature of the water from the mixer increases. It increases until it becomes saturated at 152°C. If any more steam is added the output from the mixer would be a mixture of steam and water, which could not then be compressed by the second feed pump. Therefore the best that can be done is to make the water at the outlet of the mixer just saturated. This enables the proportion of the steam bled off from the end of the high pressure turbine, y, to be found by an enthalpy balance on the mixer. Here:

$$h_1 = yh_4 + (1 - y)h_8 \qquad (14.2)$$

In the example in Fig. 14.9:

$$\begin{aligned}
h_1 &= 640 \text{ kJ/kg (saturated liquid at 5 bar)}; \\
h_4 &= 2800 \text{ kJ/kg (superheated steam at 5 bar, } 170°C); \\
h_8 &\approx h_7 = 121 \text{ kJ/kg (saturated liquid at 0.04 bar)}.
\end{aligned}$$

so that:

$$y = 0.194$$

Ignoring the work required by the feed pumps, the thermal efficiency of the cycle is:

$$\eta_{th} = \frac{(h_3 - h_4) + (1 - y)(h_5 - h_6)}{(h_3 - h_2) + (1 - y)(h_5 - h_4)} \qquad (14.3)$$

With the values given in Fig. 14.9 this efficiency is actually 0.471. From Table 14.1, it will be seen that feed heating has increased the efficiency from 0.454 to 0.471. This efficiency gain has been obtained by a relatively simple alteration to the cycle: it does not involve, for example, adding an extra expensive turbine to the cycle, just a mixer and an extra (fairly inexpensive) feed pump. Figure 14.10 shows a temperature–entropy diagram for the regenerative cycle. Note that the mixing of hot steam and cold water is a very irreversible process and therefore a source of inefficiency in the cycle. Actual cycles bleed off steam at a number of different temperatures and pressures and

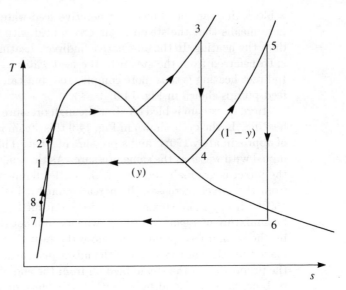

Fig. 14.10   Direct regenerative feed-water heating: temperature–entropy diagram

use these bleed streams in turn to heat the feed water to higher and higher temperatures, minimizing the temperature difference between the hot steam and the water it is heating.

14.5 Problems

14.1 In a steam–water Rankine cycle the boiler produces saturated steam at 50 bar. The condenser pressure is 0.04 bar, and the turbine is isentropic. Ignoring the feed pump work, calculate the thermal efficiency, the steam condition at the condenser inlet, and the specific steam consumption. The specific steam consumption is the mass of steam necessary to produce unit amount of work (kg/kJ).

14.2 The significance of the thermal efficiency is obvious. Comment on the significance of the specific steam consumption when comparing various cycles.

14.3 In a steam–water Rankine cycle the boiler produces saturated steam at 50 bar. The condenser pressure is 0.04 bar, and the turbine isentropic efficiency is 90%. Ignoring the feed pump work, calculate the thermal efficiency, the steam condition at the condenser inlet, and the specific steam consumption.

14.4 In a steam–water Rankine cycle the boiler produces superheated steam at 50 bar and 500°C. The condenser pressure is 0.04 bar, and the turbine isentropic efficiency is 90%. Ignoring the feed pump work, calculate the thermal efficiency, the steam condition at the condenser inlet, and the specific steam consumption.

14.5 In a steam–water Rankine cycle the boiler produces superheated steam at 50 bar and 500°C. After expansion to 5 bar, the steam is reheated to 500°C as in Figs. 14.6 and 14.7. The condenser pressure is 0.04 bar, and the isentropic efficiency of both high-pressure and low-pressure turbines is 90%. Ignoring the feed pump work, calculate the thermal efficiency, the steam condition at the condenser inlet, and the specific steam consumption.

14.6 In a steam–water Rankine cycle the boiler produces superheated steam at 50 bar and 500°C. After expansion to 5 bar, some of the steam is reheated to 500°C and some of the steam is used to heat the feed water in a regenerative feed-water heater as in Figs. 14.9 and 14.10. The condenser pressure is 0.04 bar, and the isentropic efficiency of both high-pressure and low-pressure turbines is 90%. Ignoring the feed pump work, calculate the thermal efficiency, the steam condition at the condenser inlet, the specific steam consumption, and the proportion of the steam bled off to heat the water.

14.7 Comment on the trends in the thermal efficiency and the specific steam consumption shown in the answers to questions 14.1, 14.3, 14.4, 14.5, and 14.6.

15

Steady-flow power cycles: the gas turbine cycle

15.1 Key points of this chapter

The gas turbine cycle described here is also known as the Joule cycle or the Brayton cycle.

- An open-cycle gas turbine can be analyzed on the basis of the air equivalent circuit. (Section 15.2)

- The gas turbine cycle can have high efficiency, but because the work ratio is low it is sensitive to the effect of irreversibilities in the compressor and the turbine. (Section 15.2)

- There is an optimum pressure ratio which maximizes the work output per unit mass of gas flowing. (Section 15.2)

- The gas turbine cycle can be improved, in some cases, by adding a heat exchanger to transfer heat from the turbine exit to the compressor exit. (Section 15.3)

- The effect on the gas turbine cycle of compressing in a number of stages (with cooling between the stages), and of expanding in a number of stages (with re-heating between the stages) is discussed. In general this does not improve the efficiency. (Section 15.3)

- These improvement methods, although attractive in theory, involve both heavy and expensive heat exchangers operating with gas. (Section 15.3)

- A combination of a gas turbine cycle and a steam–water Rankine cycle can be used to attain very high efficiencies. (Section 15.4)

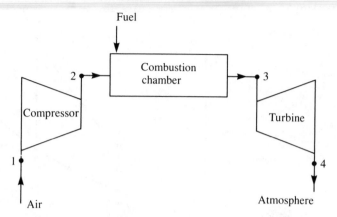

Fig. 15.1 Simple block diagram of gas turbine cycle

15.2 The basic gas turbine cycle

A gas turbine cycle is shown in block diagram form in Fig. 15.1. Strictly this is not a cycle. Air from the atmosphere is compressed, mixed with fuel and burnt in a combustion chamber, and then expanded in a turbine. The remaining gases, the products of the combustion, are returned to the atmosphere. This is often known as an 'open-cycle gas turbine'. The function of the combustion is merely to increase the temperature of the gas. Gas turbine cycles are usually analyzed on the basis of the 'air-standard cycle'. Here the working fluid, the gas, is assumed to have the properties of air (and moreover the air is assumed to behave ideally) at all points in the cycle. The increase in mass flow due to the added fuel is also neglected in this simple type of analysis. The cycle behaves as if the gas returned to the atmosphere were actually cooled before the beginning of the cycle again.

The temperature–entropy diagram for the gas turbine is shown in Fig. 15.2. From Chapter 13 the temperature ratio across the compressor and turbine, if they are assumed isentropic, is:

$$\frac{T_2}{T_1} = \frac{T_3}{T_4} = p_{\mathrm{r}}^{\frac{\gamma-1}{\gamma}} \qquad (15.1)$$

where $p_{\mathrm{r}} = p_2/p_1$ is the pressure ratio across the compressor and the turbine. Because all the heat and work transfers are proportional to temperature differences, the thermal efficiency can be written as:

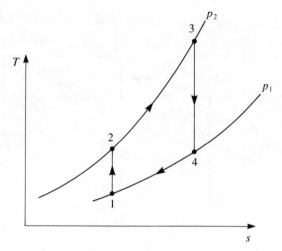

Fig. 15.2   Gas turbine cycle: temperature–entropy diagram

$$\eta_{\text{th}} = \frac{\text{net work out}}{\text{heat input}} = \frac{(T_3 - T_4) - (T_2 - T_1)}{(T_3 - T_2)} \tag{15.2}$$

Substituting from eqn 15.1 into eqn 15.2, the efficiency can be simplified to:

$$\eta_{\text{th}} = 1 - \frac{1}{p_r^{\frac{\gamma-1}{\gamma}}} \tag{15.3}$$

The variation of thermal efficiency with pressure ratio is shown in Fig. 15.3 taking $\gamma = 1.4$: apparently the higher the pressure ratio the better. However the work ratio r_w (the ratio of net work to positive work) should also be considered. Again this is a ratio of temperature differences:

$$r_w = \frac{\text{net work out}}{\text{work out}} = \frac{(T_3 - T_4) - (T_2 - T_1)}{(T_3 - T_4)} \tag{15.4}$$

or, using eqn 15.1:

$$r_w = 1 - \frac{T_1}{T_3} p_r^{\frac{\gamma-1}{\gamma}} \tag{15.5}$$

The maximum possible sensible pressure ratio occurs when the work ratio, from eqn 15.5, is equal to 0. This occurs when:

$$p_{r,\text{max}} = \left(\frac{T_3}{T_1}\right)^{\frac{\gamma}{\gamma-1}} \tag{15.6}$$

Equation 15.5 is also plotted in Fig. 15.3 for the particular case where $T_1 = 300$ K and $T_3 = 1000$ K. At a pressure ratio of 67.6, given by

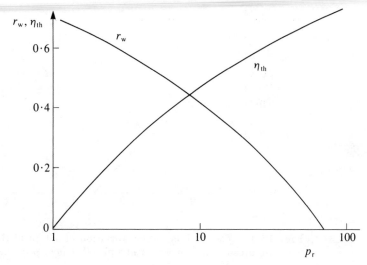

Fig. 15.3 Gas turbine cycle: variation of thermal efficiency and work ratio with pressure ratio: $T_3 = 1000$ K, $T_1 = 300$ K

eqn 15.6, the work ratio is zero, and so the cycle is producing no net work. What is happening here is that at this pressure ratio $T_3 = T_2$ and so the compressor work and the turbine work are equal, and no heat is input in the combustion chamber. Notice that, although at moderate values of the pressure ratio—around 5 to 10—the efficiency is quite high, the work ratio is small. This implies that the gas turbine cycle is sensitive to the actual efficiencies of the turbine and the compressor. Repeating the calculation of overall thermal efficiency using values for the isentropic efficiency of the compressor and the turbine η_C and η_T using the methods from Chapter 13 is not difficult. Figure 15.4 shows a typical variation of thermal efficiency for the cycle with values for the isentropic efficiency of the compressor and turbine (here assumed equal).

Note how here the overall thermal efficiency is very sensitive to the overall efficiency of the individual components. The corresponding figure for the Rankine cycle in Chapter 14 showed a much smaller variation of thermal efficiency. This is because the Rankine cycle has a large work ratio.

Another parameter of interest is the specific work output, W_{swo}. This is the net work output per unit mass of fluid flowing. In this case:

$$W_{swo} = c_p[(T_3 - T_4) - (T_2 - T_1)] \qquad (15.7)$$

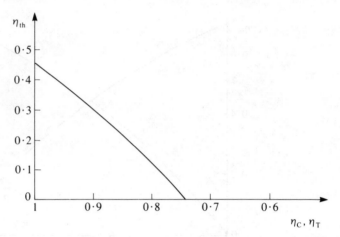

Fig. 15.4 Gas turbine cycle: variation of a typical thermal efficiency with compressor and turbine isentropic efficiency: $p_r = 8.2$

Substituting from eqn 15.1 and differentiating W_{swo} with respect to pressure ratio, the specific work output is found to be a maximum when:

$$p_r = \left(\frac{T_3}{T_1}\right)^{\frac{\gamma}{2(\gamma - 1)}} \tag{15.8}$$

or, from eqn 15.6 when:

$$p_r = \sqrt{p_{r,\max}} \tag{15.9}$$

Gas turbine cycles tend to be operated with pressure ratios around this value. Larger pressure ratios do give larger thermal efficiencies (see Fig. 15.3) but the gain in efficiency at higher pressure ratios is comparatively small. The most common application of gas turbines is in jet aircraft engines. Here it is important for weight reasons to maximize the work for given values of the inlet temperature T_1 and the temperature at the inlet to the turbine T_3.

15.3 Improvements to the gas turbine cycle

Just as there various improvements that can be made to the Rankine cycle, there are also a range of improvements that can be made to the gas turbine cycle. Again the improvements are aimed at making the cycle more like a Carnot cycle, increasing the average temperature at

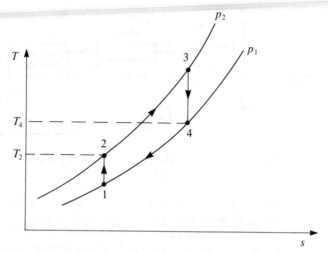

Fig. 15.5   Heat exchange in a gas turbine cycle

which heat is added to the cycle, and decreasing the average temperature at which heat is removed from the cycle. It is generally assumed that there is a limit to the maximum temperature in the cycle (the turbine entry temperature) which is fixed by the material properties of the turbine blades. There are various ways in which, in recent years, the turbine entry temperature, for example in aircraft engines, has been increased. These include internally cooling the blades in the turbine, and also externally cooling them by feeding some cool gas into the flow next to the blade through holes on the blade surface.

Two main improvements to the cycle are considered here. Neither are relevant to aircraft engines because both involve the use of heat exchangers. Heat exchangers for gas flows are always large, and therefore heavy: this precludes their use on aircraft. The improvements are heat exchange between the turbine exhaust gas and the compressor outlet gas, and the use of two turbines and re-heating the gas between them which is in itself not beneficial but can increase the scope for heat exchange.

First the use of heat exchange will be considered by reference to the cycle shown in Fig. 15.5. It is evident that in this cycle the temperature of the gas at the exit of the turbine, T_4, is greater than the temperature at the exit of the compressor, T_2. Hot gas is being thrown away to the atmosphere whilst expensive fuel is being used to heat gas up from a lower temperature. The remedy is to insert a heat exchanger into the circuit so that the compressor outlet gas is

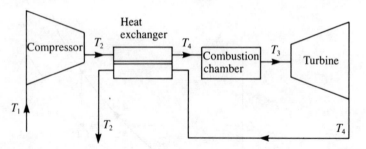

Fig. 15.6   Heat exchange in a gas turbine cycle: block diagram of circuit

heated by the turbine outlet gas (see Fig. 15.6). The mass flows of gas at these points are assumed to be equal[1], and the specific heats of the two gases are also assumed equal. With these assumptions a perfect heat exchanger will, if the streams are flowing in opposite directions as shown in Fig. 15.6, 'counter-current' operation, raise the temperature of the compressor outlet gas from T_2 to T_4, and reduce the temperature of the turbine outlet gas from T_4 to T_2 as shown in Fig. 15.5. This has the effect of reducing the amount of heat added to the cycle, and so changes the thermal efficiency from eqn 15.2 to:

$$\eta_{\text{th}} = \frac{(T_3 - T_4) - (T_2 - T_1)}{(T_3 - T_4)} \tag{15.10}$$

Comparing this equation with eqn 15.4 for work ratio it will be seen that the equations are identical, and so when heat exchange is used the thermal efficiency is given by eqn 15.5. However, heat exchange is not always possible. It is necessary for T_4 to be greater than T_2. For pressure ratios less than $\sqrt{p_{\text{r,max}}}$ this is true, and so heat exchange can be used. When the pressure ratio is greater than $\sqrt{p_{\text{r,max}}}$ it is not true, and so heat exchange cannot be used. Taking this into account the complete variation of thermal efficiency with pressure ratio, making use of heat exchange where possible, is shown in Fig. 15.7.

From the results of Chapter 13, the advantages of inter-cooling and re-heating seem to follow logically. The compressor work can be reduced using inter-cooling, and in a similar way the turbine work can be increased. Figure 15.8 shows the temperature–entropy diagram. From Chapter 13 it will be recalled that the optimum intermediate pressure p_i is equal to $\sqrt{p_1 p_2}$. However, it is not true that the

---

[1] This cannot actually be correct as the mass of the fuel added in the combustion chamber should be taken into account.

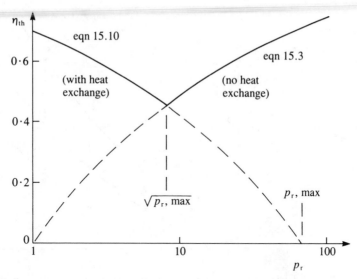

Fig. 15.7 Variation of thermal efficiency with pressure ratio, heat exchange used where possible

addition of inter-cooling and re-heat increase the overall thermal efficiency. The work of compression is certainly reduced, and the turbine work output is certainly increased, but unfortunately the heat input has increased significantly so that the resulting efficiency is actually reduced. This can be appreciated from the temperature–entropy diagram in Fig. 15.8. The basic cycle in this figure is 12′56′. To this basic cycle two subsidiary cycles are have been added by the inter-cooling (2342′), and by the re-heating (6786′). The two subsidiary cycles have low efficiency since they operate over a small temperature range. Thus their addition to the main cycle lowers the cycle thermal efficiency. However, re-heating can be advantageous, not because by itself it increases the efficiency, but the final turbine outlet temperature is increased, and so the scope for heat exchange is increased. This point is illustrated in the following example.

Example 15.1   A simple gas turbine cycle has a pressure ratio of 9, an inlet temperature of 300 K, and a turbine inlet temperature of 1000 K. The gas has the properties of air. Find the thermal efficiency of the cycle. Can heat exchange be used to increase the efficiency? Now two turbines are used with re-heating back to 1000 K between them. The individual pressure ratio for each turbine is 3. Calculate the thermal efficiency. Can heat exchange now be used to increase the efficiency?

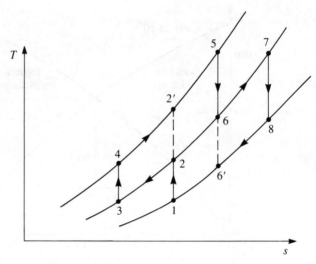

Fig.   15.8   Gas   turbine   cycle   with   inter-cooling   and   re-heat:
temperature–entropy diagram

Solution   The simple cycle with a pressure ratio of 9 is that shown in Fig. 15.1,
and in Fig. 15.2. Working out the temperatures in Fig. 15.2, from
eqn 15.1:

$$T_2 = T_1 p_r^{\frac{\gamma-1}{\gamma}} = 300 \times 9^{\frac{1.4-1}{1.4}} = 562 \text{ K}$$

$$T_4 = T_3/p_r^{\frac{\gamma-1}{\gamma}} = 1000/9^{\frac{1.4-1}{1.4}} = 534 \text{ K}$$

In this case T_4 is less than T_2 and so heat exchange is not possible.
The thermal efficiency from eqn 15.2 or eqn 15.3 is 0.466. The cycle
with re-heat but no inter-cooling is the cycle 1234561 in Fig. 15.9.
Again $T_2 = 562$ K. T_3 and T_5 are both 1000 K. T_4 and T_6 are equal,
and are given by:

$$T_4 = T_6 = T_3/p_r^{\frac{\gamma-1}{\gamma}} = 1000/3^{\frac{1.4-1}{1.4}} = 731 \text{ K}$$

The efficiency of the cycle is:

$$\eta_{th} = \frac{(T_3 - T_4) + (T_5 - T_6) - (T_2 - T_1)}{(T_3 - T_2) + (T_5 - T_4)} = 0.390$$

The addition of the re-heating has decreased the efficiency from 0.466
to 0.390. However, heat exchange is now possible as T_6 is larger than

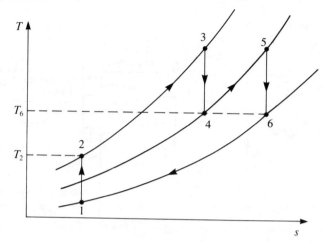

Fig. 15.9 Example 15.1: gas turbine cycle with and re-heat and heat exchange: temperature–entropy diagram

T_2. With perfect heat exchange the gas at the outlet of the compressor can be heated from T_2 to T_6, and so the efficiency becomes:

$$\eta_{th} = \frac{(T_3 - T_4) + (T_5 - T_6) - (T_2 - T_1)}{(T_3 - T_6) + (T_5 - T_4)} = 0.513$$

The combination of re-heat and heat exchange, as in Fig. 15.10, has increased the efficiency from 0.466 to 0.513.

15.4 Combined cycles: gas turbine and Rankine cycles

The example at the end of the previous section has a very high efficiency, but even larger efficiencies are possible. The problem with the re-heat and heat exchange cycle is that the heat exchanger is large and expensive. Both the streams in the heat exchanger are gas, and heat transfer to or from gas is relatively poor. That both streams are gas makes the problem very severe. Also even with heat exchange, the gas is rejected to the atmosphere at the relatively high temperature of 562 K (289°C). This temperature can be reduced, and the heat exchange can be improved by using a combination of a gas turbine cycle and a Rankine cycle (see Fig. 15.11). Here the hot exhaust gas at the end of the second gas turbine is used to heat water to its saturation temperature, vaporize the water, and superheat the steam. This

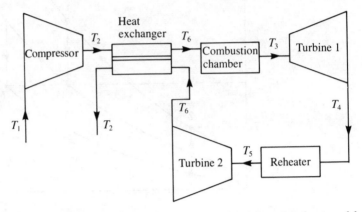

Fig. 15.10 Example 15.1: gas turbine cycle with and re-heat and heat exchange: block diagram

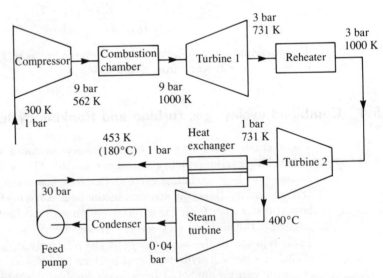

Fig. 15.11 Combined gas turbine and Rankine cycle: block diagram

steam is then expanded in a steam turbine. The heat exchanger now has water as one of its streams, and is therefore significantly smaller than a gas–gas heat exchanger. The temperature of the gas at the outlet of the heat exchanger cannot be too low. If the temperature is reduced too far, the water vapour in the gas produced inevitably as the hydrocarbon[2] fuel burns condenses to form liquid water. Any sulphur in the fuel makes this water acidic and corrosive. For this reason it is unwise to reduce the gas temperature below around 180°C.

The detailed analysis of this combined cycle is outside the scope of this book, but the efficiency of the combined cycle can be greater than the gas turbine cycle with re-heat and heat exchange. For example, with the data in Fig. 15.11 the overall efficiency of the cycle is 0.520. One of the more difficult decisions in designing these combined cycles is deciding the pressure of the water cycle (here chosen as 30 bar, corresponding to a saturation temperature of 234°C). Too low a pressure gives a poor overall thermal efficiency, and too high a pressure will give a region in the heat exchanger where the temperature difference between the hot gas and the colder water side is too low.

15.5 Problems

15.1 A gas turbine cycle has an isentropic compressor and an isentropic turbine. How does the thermal efficiency vary with γ over the realistic range for γ? For sample calculations assume a pressure ratio, p_r of 8.

15.2 Air at 1 bar and 300 K is compressed isentropically to 6 bar. It is then heated at constant pressure to 1000 K, and expanded reversibly and adiabatically in a turbine to 1 bar. Calculate the temperature at the end of the compression, and at the end of the expansion. Hence calculate the net work output per kg of gas, and the heat input per kg of gas. From these quantities calculate the thermal efficiency, and verify that your answer is consistent with the result obtained from eqn 15.3.

15.3 Air at 1 bar and 300 K is compressed to 6 bar with an isentropic efficiency of 85%. It is then heated at constant pressure to 1000 K, and expanded in a turbine to 1 bar with an isentropic efficiency of 90%. Calculate the temperature at the end of the compression, and at the end of the expansion. Hence calculate the net work output per kg of gas, and the heat input per kg of gas. From these quantities calculate the thermal efficiency.

[2] A hydrocarbon molecule contains carbon and hydrogen atoms. On combustion the hydrogen atoms form water vapour, and the carbon atoms form carbon dioxide.

15.4 For the cycles in questions 15.2 and 15.3 can heat exchange be used to increase the thermal efficiency? If so, what are the resulting efficiencies?

15.5 In the combined gas turbine and Rankine cycle shown in Fig. 15.11 all the compressors, pumps, and turbines are isentropic. If the gas has the properties of air:

(a) verify the gas side temperatures shown in Fig. 15.11 up to the end of the second turbine;

(b) assuming the gas leaves at 180°C as shown, calculate the heat removed per kg of gas;

(c) calculate the heat added to the water per kg of water flowing;

(d) hence calculate m_r, the ratio of the gas mass flow rate to the water mass flow rate;

(e) verify that the thermal efficiency of the combined cycle is 52.0%.

15.6 Using the data and results from question 15.5 plot, on one diagram, the enthalpy of the streams versus the temperature. Plot the enthalpy per kg of water, and per m_r kg of air. Hence find the minimum temperature difference between the two streams along the heat exchanger. Hence explain why the water side pressure could not be very much greater than 30 bar.

16

Heat pump and refrigeration cycles

16.1 Key points of this chapter

- The most efficient heat pump or refrigerator is a Carnot cycle. However, this is difficult to achieve in practice. (Section 16.2)

- A more practical cycle is one based on the compression of a saturated vapour, its condensation, a decrease in the pressure, followed by evaporation of liquid. (Section 16.3)

- For practical reasons a throttle valve is usually used to let the pressure down, even though the use of a turbine would be more efficient. (Section 16.4)

- For non-ideal cycles the use of a pressure–enthalpy diagram can be very useful. (Section 16.5)

- Another refrigeration cycle which works on an entirely different principle is the absorbtion cycle. (Section 16.6)

16.2 The Carnot cycle used as a refrigerator or a heat pump

Just as it was proved in Chapter 9 that the reversible heat engine (that is, the Carnot cycle) is the most efficient heat engine operating between two temperature reservoirs, a Carnot cycle is also the most efficient heat pump or refrigerator. From Chapter 8 where the definition of coefficient of performance is given, and Chapter 9 where the efficiency of the Carnot cycle is given, it is easy to prove the following.

1. For a refrigerator, the Carnot coefficient of performance is:

$$(COP)_R = \frac{\text{heat removed from cold reservoir}}{\text{work required}}$$

$$= \frac{T_{\text{low}}}{T_{\text{high}} - T_{\text{low}}} \tag{16.1}$$

2. For a heat pump, the Carnot coefficient of performance is:

$$(COP)_{HP} = \frac{\text{heat delivered to hot reservoir}}{\text{work required}}$$

$$= \frac{T_{\text{high}}}{T_{\text{high}} - T_{\text{low}}} \tag{16.2}$$

Thus for temperatures of:

$$T_{\text{high}} = 10°C = 283 \text{ K}$$

and:

$$T_{\text{low}} = -35°C = 238 \text{ K}$$

the values of the coefficients of performance are:

$$(COP)_R = 5.29 \quad \text{and} \quad (COP)_{HP} = 6.29$$

For these temperatures, these values are the largest coefficients of performance that can be obtained. However, just as it is not practical to have a large-scale Carnot heat engine, it is also not practical to have a large-scale Carnot refrigerator. For example, it is advantageous to use a vapour–liquid system because the latent heat of evaporation is large and so the heat transfers can be quite large without having large flows of fluid. Figure 16.1 shows a Carnot cycle superimposed on the saturated vapour–liquid line on a temperature–entropy diagram. In this cycle the isothermal processes can easily be carried out: they are, after all, simply an evaporation and a condensation. The saturated liquid can, theoretically, be expanded in an isentropic turbine to give a mixture of liquid and vapour, but it is the compression of a vapour–liquid mixture to give saturated vapour which presents the real difficulty. Vapours, just like gases, can be compressed. Liquids also can be compressed, though a liquid compressor is usually thought of as a pump. However, machines to compress vapour–liquid mixtures satisfactorily do not exist. Hence this Carnot cycle shown in Fig. 16.1 cannot be achieved in practice.

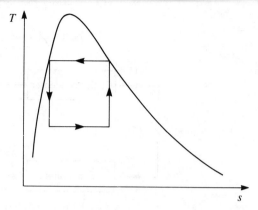

Fig. 16.1   Carnot refrigerator with a vapour–liquid system

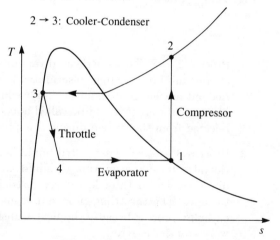

Fig. 16.2   Practical refrigerator with a vapour–liquid system: temperature–entropy diagram

16.3   The practical cycle for a refrigerator or a heat pump

Figure 16.2 shows the temperature–entropy diagram of a practical refrigeration process. A block diagram of the process is shown in Fig. 16.3. The main features of he cycle are as follows.

1. Process 1 → 2. Saturated vapour is compressed to give super-heated vapour.

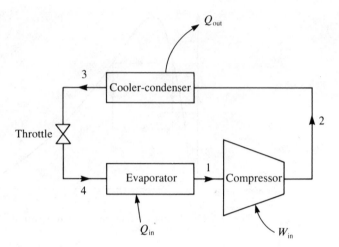

Fig. 16.3   Practical refrigerator with a vapour–liquid system: block diagram

2. Process $2 \to 3$. The superheated vapour is cooled and condensed at constant pressure to give saturated liquid. This process is carried out in what is often misleadingly simply called a 'condenser'. More accurately it would be called a 'cooler–condenser'. Heat is rejected from the system during this process.

3. Process $3 \to 4$. The saturated liquid is passed through a throttle valve to give a mixture of liquid and vapour. Section 16.4 discusses why a throttle valve is used here in preference to a turbine. The throttling process is usually taken as a constant enthalpy process because the heat transfer is negligible and the work transfer is zero.

4. Process $4 \to 1$. The mixture of liquid and vapour is evaporated until there is just saturated vapour present. Heat is taken into the system during this process.

An example makes use of the carbon dioxide compressor used as an example in Chapter 13.

Example 16.1   A heat pump–refrigerator cycle uses carbon dioxide as the working fluid. The high and low pressures are 45 bar and 12 bar where the saturation temperatures are 10°C and -35°C. The compressor isentropic efficiency is 85%. Find the coefficient of performance for both the refrigerator and the heat pump.

Solution   From the property values in Chapter 13, the values of the enthalpy at various points in the cycle are as follows (see Figs. 16.2 and 16.3):

Point 1   saturated vapour at 12 bar $h_1 = 322.2$ kJ/kg
Point 2   superheated vapour at 45 bar $h_2 = 391.5$ kJ/kg
(from Chapter 13)
Point 3   saturated liquid at 45 bar $h_3 = 109.1$ kJ/kg
Point 4   liquid and vapour at 12 bar $h_4 = 109.1$ kJ/kg

Note that $h_3 = h_4$. Now:

$$(\text{COP})_\text{R} = \frac{\text{heat removed from cold reservoir}}{\text{work required}} = \frac{h_1 - h_4}{h_2 - h_1} \quad (16.3)$$

$$(\text{COP})_\text{HP} = \frac{\text{heat delivered to hot reservoir}}{\text{work required}} = \frac{h_2 - h_3}{h_2 - h_1} \quad (16.4)$$

and substituting values for the enthalpies gives:

$$(\text{COP})_\text{R} = 3.08 \quad \text{and} \quad (\text{COP})_\text{HP} = 4.08$$

Clearly these values for the coefficient of performance are significantly lower than the values for a Carnot cycle operating between -35°C and 10°C quoted in Section 16.2. Three reasons are immediately obvious to explain this difference.

1. The compressor is not isentropic, and therefore requires a greater work input than an isentropic compressor.

2. The heat is not all liberated at 10°C. Some of the heat is liberated at a higher temperature, contributing to a lower coefficient of performance.

3. Work is wasted in the throttle. The significance of this is examined in the Section 16.4.

Carbon dioxide is used as a refrigerant only in very large systems. It is used particularly in refrigerated ships. For smaller-scale systems a refrigerant is used which, in the temperature range of interest, has a smaller saturation pressure than carbon dioxide. Pressures in the region of 45 bar are inconveniently large for domestic operation.

The most common domestic refrigerant in use today is dichloro-difluoro-methane, with the chemical formula CCl_2F_2[1]. At -35°C the

---

[1] Dichloro-difluoro-methane is also known as Refrigerant-12 (R-12), Freon-12, and Arcton-12. Freon is a trade name of DuPont, and Arcton is a trade name of ICI.

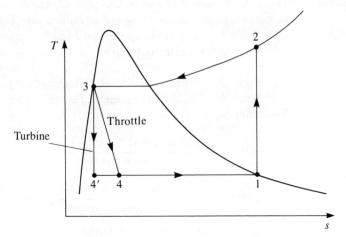

Fig. 16.4 Practical refrigerator with a vapour–liquid system using a turbine to expand the liquid: temperature–entropy diagram

saturation pressure is 0.8 bar, and at 10°C it is 4.2 bar. In recent years environmental pressures have meant that substitutes for CCl_2F_2 are being found. It is probable that the substitutes will not have such suitable thermodynamic properties, and will be much more expensive.

16.4 A turbine or a throttle for the expansion process?

Use of a turbine to expand the saturated high-pressure liquid rather than a throttle will produce some extra work. This will reduce the overall work requirements of the cycle and so increase the coefficient of performance. This desirable objective is not followed in practice for a number of reasons.

1. The throttle valve is cheap and reliable. The turbine would be expensive, and provide more moving parts in the refrigerator which will decrease the reliability. It is worth noting the extraordinary reliability of domestic-type refrigerators. Often they run with no maintenance for ten years. Much effort has gone into the design of the compressor and the electric motor to power it. These are built as integral sealed units.

2. The amount of work that can be produced is quite small. The temperature–entropy diagram of the cycle with an isentropic tur-

bine instead of a throttle valve is shown in Fig. 16.4. In this diagram points 1, 2, and 3 are unchanged from Fig. 16.2. Only the enthalpy of point 4' need be recalculated. This point is at a pressure of 12 bar, and the entropy is the same as for point 3. Hence using the data from Chapter 13:

$$s_3 = s_{4'} = (1 - x_{4'})s_{f4'} + x_{4'}s_{g4'} \tag{16.5}$$

Substituting into this equation:

$$
\begin{aligned}
s_3 = s_{4'} &= 0.407 \text{ kJ/kg K} \\
s_{f4'} &= 0.039 \text{ kJ/kg K} \\
s_{g4'} &= 1.352 \text{ kJ/kg K}
\end{aligned}
$$

gives:

$$x_{4'} = 0.280$$

Similarly:

$$h_{4'} = (1 - x_{4'})h_{f4'} + x_{4'}h_{g4'} \tag{16.6}$$

Substituting into this equation:

$$
\begin{aligned}
h_{f4'} &= 9.7 \text{ kJ/kg} \\
h_{g4'} &= 322.2 \text{ kJ/kg}
\end{aligned}
$$

gives:

$$h_{4'} = 97.3 \text{ kJ/kg}$$

The work therefore produced in the turbine is $h_3 - h_{4'}$ or $109.1 - 97.3 = 11.8$ kJ/kg. This reduces the net work required by the cycle from the previous value of 69.3 kJ/kg to 57.5 kJ/kg, and so increases the coefficients of performance by about 20%.

3. This increase in coefficient of performance is the maximum which could be obtained. In fact the efficiency of a turbine operating under these conditions is likely to be low, and so the actual increase in coefficient of performance would be significantly less than 20%.

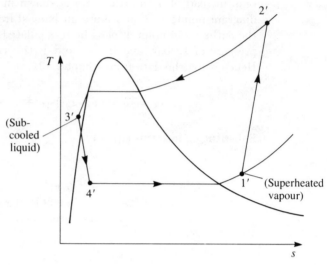

Fig. 16.5   Practical refrigerator: usual form of the temperature–entropy diagram

16.5   Use of the pressure–enthalpy diagram

Often the temperature–entropy diagram of a real refrigerator differs from that shown in Fig. 16.2. A more realistic diagram is shown in Fig. 16.5. Here the evaporation has proceeded a little too far, and the output of the evaporator is slightly superheated vapour. The condensation process has also gone a little too far, and the output of the condenser is subcooled liquid. On a pressure–enthalpy diagram these changes can be handled more easily. The cycle is shown in Fig. 16.6 as cycle $1'2'3'4'$. The simple cycle corresponding to Fig. 16.2 with no superheating at the evaporator outlet and no subcooling at the condenser outlet is cycle 1234. It can be seen that with a pressure–enthalpy diagram, superheating and subcooling present no real complications at all.

A difficulty sometimes encountered is the determination of the enthalpy of subcooled liquid without specific tabular data for subcooled liquid. It should be remembered that enthalpy, like all thermodynamic properties, is a function of two other variables, such as temperature and pressure. However, for the liquid phase enthalpy is very dependent on temperature and hardly dependent on pressure at all. Therefore a crude procedure which introduces little error into

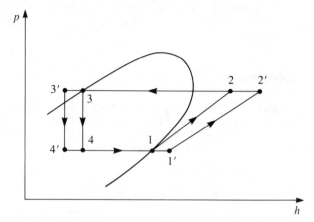

Fig. 16.6   Refrigerator cycles on a pressure–enthalpy diagram

refrigerator calculations is to look up the liquid enthalpy in the saturation table at the actual temperature, and to use this value. So to find the enthalpy of subcooled liquid carbon dioxide at 51 bar[2] and 10°C, the liquid saturation enthalpy at 10°C should be used, that is 109.1 kJ/kg. Use of the liquid saturation enthalpy at 51 bar, that is 123.3 kJ/kg, is not correct.

16.6   The absorbtion-cycle refrigerator

The refrigerator cycle previously described is known for obvious reasons as the vapour compression cycle. Some refrigerators work on an entirely different principle: these are absorbtion-cycle[3] refrigerators. This is a complicated cycle containing ammonia, hydrogen, and liquid water, and is in fact three cycles in one. The analysis of the cycle is difficult and will not be attempted here. The principle of operation is that in the presence of an inert gas (here hydrogen), some evaporation of a liquid such as liquid ammonia will take place producing a cooling effect. The gaseous ammonia is later absorbed in water, liberating heat. Circulation round the circuit is induced by density differences induced by heating part of the circuit.

The important point about the absorbtion cycle is that there is no work input. The energy input to the cycle is a heat input, and

---

[2] At this pressure the saturation temperature is 15°C.

[3] The absorbtion cycle is sometimes known as the Electrolux cycle.

therefore this type of refrigerator can be powered by burning gas. The refrigerator itself has no moving parts whatsoever and so is very reliable.

16.7   Problems

16.1 The Carnot cycle in Fig. 9.6 operates in the direction DCBAD as a refrigerator, and has air as the working fluid. The pressures at points A, C, and D are respectively: 1 bar, 3 bar, and 2 bar. The temperature at A is 250 K. Calculate:

(a) the pressure at B,

(b) the coefficient of performance for this refrigerator,

(c) the heat removed per kg of air,

(d) the diameter of the pipe required to carry the air flow at A if the heat is removed at the rate of 1 kW and the air velocity at A is limited to 10 m/s.

16.2 A refrigerator operating with the cycle shown in Figs. 16.2 and 16.3 has water as the working fluid. The low pressure in the cycle is 0.04 bar, and the high pressure is 0.50 bar. The isentropic efficiency of the compressor is 90%. Calculate:

(a) the coefficient of performance for this refrigerator,

(b) the heat removed per kg of water flowing,

(c) the diameter of the pipe required to carry the steam flow at the evaporator exit the heat is removed at the rate of 1 kW and the vapour velocity at this point is limited to 10 m/s.

16.3 A refrigerator operating with the cycle shown in Figs. 16.2 and 16.3 has ammonia as the working fluid. The low pressure in the cycle is 1.9 bar, and the high pressure is 10.0 bar. The isentropic efficiency of the compressor is 90%. Note: see question 13.8 for a relevant calculation on the compressor in this cycle. Calculate:

(a) the coefficient of performance for this refrigerator,

(b) the heat removed per kg of ammonia flowing,

(c) the diameter of the pipe required to carry the ammonia vapour flow at the evaporator exit the heat is removed at the rate of 1 kW and the vapour velocity at this point is limited to 10 m/s.

16.4 Referring to the results of questions 16.1, 16.2, and 16.3:

(a) Why is the heat removed so low for the air refrigerator in question 16.1?

(b) Why is this fact important as all three cycles have similar values of the coefficient of performance?

(c) Why is the tube diameter so small in the ammonia refrigerator in question 16.3?

(d) List the desirable properties of a good working fluid for a refrigerator.

(e) Refrigerants such as CCl_2F_2 are now regarded as environmentally unacceptable, and alternatives are being developed. What is wrong with simple alternatives like ammonia and carbon dioxide?

17

Internal combustion engine cycles

17.1   Key points of this chapter

This chapter gives only a very brief introduction to internal combustion engines, and in particular concentrates on the analysis of the cycles using idealized 'air standard' cycles.

- The four-stroke spark ignition engine is described in outline, together with the form of the indicator diagram. The cycle can only be analyzed by means of a similar but not identical cycle: the Otto air standard cycle. (Section 17.2)

- Equations for the efficiency of the Otto air standard cycle are developed, and the efficiency is found to depend only on the compression ratio of the engine. The compression ratio cannot be made too large or the fuel ignites spontaneously. (Section 17.3)

- The Diesel cycle can use a higher compression ratio because the fuel is sprayed into the hot gas. The gas is so hot that the fuel ignites spontaneously and no spark is necessary. (Section 17.4)

- For the same compression ratio the Otto cycle is more efficient than the Diesel cycle, but the greater compression ratio in Diesel engines in practice means that Diesel engines are actually more fuel efficient. (Section 17.4)

17.2   Brief description of a four-stroke spark ignition engine

Figure 17.1 shows one cylinder of a four-stroke spark ignition engine at a number of points in the cycle. Four-stroke means that the cycle

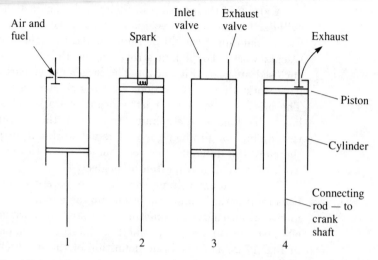

Fig. 17.1 Four-stroke spark ignition engine: various stages of the engine cycle

consists of four strokes, each stroke being a movement of the piston from one end of the cylinder to another. Thus during the cycle the crankshaft rotates through two whole revolutions. Two-stroke engines are common but are less satisfactory in operation. Spark ignition means that the mixture of fuel (usually petrol) and air is ignited by a spark produced by the spark plug. The four strokes of the engines are as follows.

1. The volume in the cylinder is increased pulling in a mixture of air and fuel which has been mixed in the carburettor.

2. The cylinder volume is decreased, compressing the mixture. Near the end of this stroke the spark ignites the mixture.

3. The mixture rapidly burns increasing the pressure and temperature in the cylinder, pushing the piston back. This is the power stroke.

4. The products of the combustion are pushed out of the cylinder as the volume again decreases. The cylinder is now ready to receive a new charge of fuel–air mixture from the carburettor.

The gases enter and leave through the inlet and outlet valves. These are shown in Fig. 17.1. Each valve opens once every two revolutions of the main engine crankshaft.

A graph of the pressure in the cylinder against the volume of the cylinder for the whole cycle is known as the indicator diagram. On a p–V diagram it should be remembered that the enclosed area represents work. For a four-stroke engine there are two parts to the enclosed area (see Fig. 17.2). The larger upper part represents positive work, that is, work done by the cylinder on the surroundings. The much smaller lower part is negative work. This is work which has to be done on the cylinder to draw in the mixture and to expel the combustion product gases. The net work output per cycle, W, is therefore the difference between these areas. However, the negative work is so small it can often be neglected.

The maximum volume change is the swept volume V_s, so-called because it is the volume swept out by the moving piston. The performance of an engine is sometimes assessed by working out the mean effective pressure, $\bar{p}_{\text{eff}}$, a geometrical interpretation of which is shown in Fig. 17.2. The algebraic definition of the mean effective pressure is:

$$\bar{p}_{\text{eff}} = \frac{W}{V_s} \tag{17.1}$$

The larger the mean effective pressure, the more work is produced during each cycle. Thus the mean effective pressure is a simple index of performance for an engine. The detailed shape of the real cycle shown in Fig. 17.2 is complicated. This is because the compression and expansion of the gas are approximately, but not of course completely, isentropic. Also the fuel burns quickly but not instantaneously. To aid our understanding of the importance of various parameters on the cycle efficiency, an approximate cycle is used. This is the Otto air standard cycle discussed below.

17.3 The Otto air standard cycle

In the Otto cycle approximation to the spark ignition engine cycle, the following approximations are made.

1. The properties of the gas are assumed to be those of air. This is a useful simplification as the properties of, in particular, the combustion products can be difficult to ascertain.

2. The negative work regions where the mixture is drawn into the cylinder and the combustion products are expelled are ignored completely.

3. The compression and expansion of the gas are assumed to be isentropic.

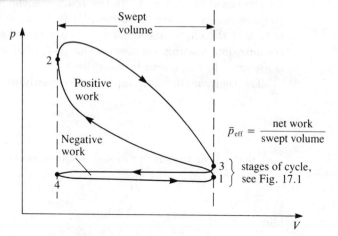

Fig. 17.2   Indicator diagram for the four-stroke spark ignition engine

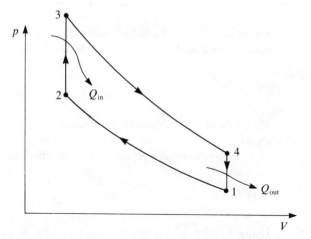

Fig. 17.3   The Otto air standard cycle

4. The combustion is assumed to occur instantaneously when the cylinder is at its minimum volume.

With these assumptions the Otto air standard cycle is shown on a p–V diagram in Fig. 17.3. The processes $1 \rightarrow 2$ and $3 \rightarrow 4$ are the isentropic processes, so these processes are adiabatic. The heat is added in the process $2 \rightarrow 3$ where the combustion occurs, and the heat is removed at $4 \rightarrow 1$. In the real cycle the gas at 4 is rejected to the atmosphere, and fresh gas at 1 is drawn in. It is important to note that the Otto air standard cycle is a closed system. During the compression, heating, expansion, and cooling processes the same gas is assumed to be present in the cylinder.

The thermal efficiency, η_{th}, is conveniently written as:

$$\eta_{th} = 1 - \frac{Q_{out}}{Q_{in}} \tag{17.2}$$

and as the heat transfer processes occur at constant volume:

$$Q_{in} = c_v(T_3 - T_2) \tag{17.3}$$

and:

$$Q_{out} = c_v(T_4 - T_1) \tag{17.4}$$

so that:

$$\eta_{th} = 1 - \frac{T_4 - T_1}{T_3 - T_2} \tag{17.5}$$

Equation 17.5 can be further simplified by remembering that the processes $1 \rightarrow 2$ and $3 \rightarrow 4$ are isentropic, so that:

$$\frac{T_2}{T_1} = \frac{T_3}{T_4} = r_v^{\gamma-1} \tag{17.6}$$

where r_v is the volumetric compression ratio of the engine. This is equal to V_1/V_2 and to V_4/V_3. Substituting eqn 17.6 into eqn 17.5 gives an even simpler equation for the efficiency:

$$\eta_{th} = 1 - \frac{1}{r_v^{\gamma-1}} \tag{17.7}$$

Using a value of 1.4 for γ, this equation is plotted in Fig. 17.4 as a function of the compression ratio. To obtain a high efficiency the compression ratio must be large, but extreme values like 20 as shown in Fig. 17.4 are not possible. During the initial isentropic compression, process $1 \rightarrow 2$, in Fig. 17.3 the gas becomes hot. If the gas is compressed too much it gets so hot it ignites spontaneously and unpredictably. This is pre-ignition and does not lead to satisfactory performance. In practice, compression ratios of around 9 to 10 are used. The addition of the chemical compound tetra-ethyl lead delays spontaneous ignition. Alternative lead-free petrol has to be more highly refined to give satisfactory performance.

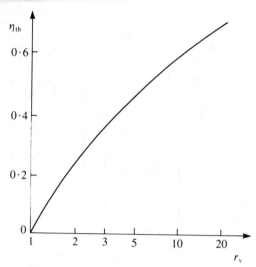

Fig. 17.4   The Otto air standard cycle: variation of efficiency with compression ratio

17.4   The Diesel cycle and its air standard equivalent

The four-stroke Diesel engine is shown in simple form in Fig. 17.5, which is rather similar to Fig. 17.1 for the spark ignition engine. The four strokes are as follows.

1. Air is pulled into the cylinder.

2. The air is compressed, and at the end of the stroke the fuel begins to be sprayed into the cylinder. It enters through a spray nozzle, a very accurately made nozzle which is designed to atomize the fuel into small droplets which will burn well. After the compression the air is hot enough to ignite the fuel.

3. The burning fuel and air is allowed to expand. Part way through the expansion the fuel spray is stopped.

4. The combustion products are pushed out of the cylinder and the cylinder is ready to receive a new charge of air.

Because the fuel is sprayed into the cylinder relatively slowly, and because the fuel is liquid, the burning of the fuel continues for a

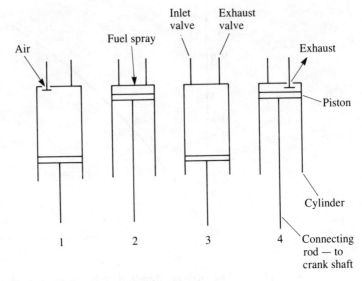

Fig. 17.5 The Diesel engine: various stages of the engine cycle

relatively long time[1]. In the Diesel engine the burning of the fuel tends to increase the pressure, whereas the increase in the volume of the cylinder as the fuel burns tends to decrease the pressure. These contrasting effects tend to give a part of the cycle where the pressure is very roughly constant. This observation is made use of in the air standard equivalent of the Diesel cycle shown in Fig. 17.6. As with the Otto cycle this is a closed system process: during the following processes the same gas is assumed to be present in the cylinder. Again there are two isentropic processes: $1 \rightarrow 2$, and $3 \rightarrow 4$. The heat is assumed to be added at constant pressure in the process $2 \rightarrow 3$, and the heat removed at constant volume in the process $4 \rightarrow 1$. Again eqns 17.2 and 17.4 are valid, but eqn 17.3 must be replaced by:

$$Q_{in} = c_p(T_3 - T_2) \qquad (17.8)$$

as the heat input process is now at constant pressure. The simple equation 17.6 for the temperature ratios must be replaced by a more complicated set of equations.

$$\frac{T_2}{T_1} = r_v^{\gamma-1} \qquad (17.9)$$

[1]In contrast in a spark ignition engine, the fuel is often vaporized before it enters the engine, and the fuel–air mixture is a vapour which burns explosively.

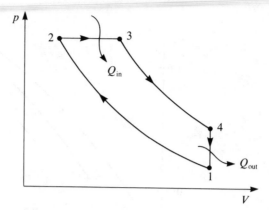

Fig. 17.6   The Diesel air standard cycle

$$\frac{T_3}{T_2} = r_c \tag{17.10}$$

and

$$\frac{T_4}{T_3} = \left(\frac{r_c}{r_v}\right)^{\gamma - 1} \tag{17.11}$$

where r_v is the compression ratio V_1/V_2, and r_c is the 'cut-off' ratio V_3/V_2.

After some manipulation it can be shown from eqns 17.2, 17.4, and 17.8 to 17.11 that:

$$\eta_{th} = 1 - \frac{1}{r_v^{\gamma - 1}}\left[\frac{r_c^{\gamma} - 1}{\gamma(r_c - 1)}\right] \tag{17.12}$$

Note that for the special case of $r_c = 1$, this efficiency reduces to that given by eqn 17.7, the Otto cycle efficiency. Figure 17.7 shows a graph of eqn 17.12 plotted against compression ratio with compression ratio for various values of the cut-off ratio, again using a value of 1.4 for γ. Note that for a given compression ratio, the Diesel cycle is less efficient than an Otto cycle, but because a Diesel engine can operate with a higher value of the compression ratio, the final efficiency of a Diesel engine is greater. To get the air hot enough to ignite the fuel the compression ratio must be at least 12. The actual compression ratio used in a Diesel engine can be up to 20.

The advantages of the Diesel engine are therefore greater efficiency and a simpler engine: there is no spark plug and no high-voltage system to go wrong. The disadvantages are that the engine is heavier as the cylinders have to contain a greater pressure, and it is relatively

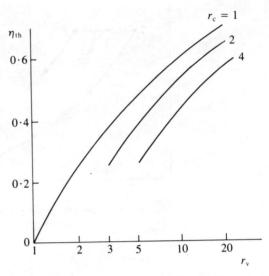

Fig. 17.7 The Diesel air standard cycle: variation of efficiency with compression ratio and cut-off ratio

expensive to manufacture (particularly the very precise machining of the fuel injection system). In spite of considerable advances over recent years Diesel engines still tend to need more maintenance than petrol engines and they are still rather noisy, particularly at low speed.

17.5 Problems

17.1 The Otto cycle is shown in Fig. 17.3. The gas behaves ideally with a value of R of 300 J/kg K and γ of 1.30. The gas at point 1 is at 1 bar and 300 K. The mass ratio of air to fuel is 14:1. The fuel has a calorific value of 42 MJ/kg. [The calorific value can be regarded as the amount of heat released when the fuel burns.] The compression ratio is 7, calculate:

(a) the temperature of the gas at points 2, 3, and 4;

(b) the pressure in the cylinder at points 2, 3, and 4;

(c) the heat added to the gas, per kg of mixture, when the gas burns;

(d) the heat rejected in the gas in the process 4 → 1, per kg of mixture;

(e) the thermal efficiency, calculated from the heat added and rejected, confirm this agrees with the efficiency calculated from eqn 17.7;

(f) the mean effective pressure for this cycle.

17.2 Explain why the calculation in question 17.1 is unrealistic, and in what important respects the real cycle will differ from that assumed.

17.3 The cycle in question 17.1 is now supercharged by compressing the mixture at point 1 before the mixture enters the cylinder. If the inlet mixture is compressed to 2 bar and cooled back to 300 K:

(a) What is the thermal efficiency of the supercharged cycle?

(b) What is the mean effective pressure of the supercharged cycle?

(c) From your results, discuss why supercharging might be used. What are the advantages and disadvantages? How might the power to run the compressor to raise the pressure of the mixture be obtained?

17.4 Prove eqn 17.12 for the efficiency of the Diesel cycle. Prove that when the cut-off ratio r_c is equal to 1, the equation reduces to eqn 17.7.

17.5 A four-stroke four-cylinder Diesel engine operating at 2000 revolutions per minute consumes 4.5 litres of fuel per hour. Calculate the volume of fuel injected into each cylinder before each power stroke. Comment on the result. If the total cylinder capacity of the engine is 2 litres, and the air is at atmospheric conditions after it has been sucked into the cylinder, calculate the air:fuel mass ratio. The fuel density is 800 kg/m^3.

18

Thermodynamic equations

18.1 Key points of this chapter

- The four Maxwell equations relating various thermodynamic variables are very useful in many thermodynamic derivations, and in the manipulation of thermodynamic equations. (Section 18.2)

- The Maxwell equations can be used to prove that, for an ideal gas, the internal energy and the enthalpy depend only on the temperature. (Section 18.3)

- The Maxwell equations can also be used to derive a general equation for the difference between c_p and c_v. For an ideal gas this general equation reduces to the expected result: $c_p - c_v = R$. (Section 18.3)

- The properties of the Gibbs function are used to derive the Clausius–Clapeyron equation which described the variation of saturation vapour pressure with temperature. (Section 18.4)

- Bernouilli's equation is derived very simply from the steady-flow energy equation. Bernouilli's equation is one of the basic equation of fluid mechanics, and describes the variation of pressure with velocity in a moving fluid. (Section 18.5)

18.2 The Maxwell equations

If, in general, we have a functional relationship of the form:

$$z = f(x, y) \tag{18.1}$$

then:

$$dz = \left(\frac{\partial z}{\partial x}\right)_y dx + \left(\frac{\partial z}{\partial y}\right)_x dy \qquad (18.2)$$

which can be written as:

$$dz = M\,dx + N\,dy \qquad (18.3)$$

Now from the theory of partial differentiation:

$$\frac{\partial^2 z}{\partial x \partial y} = \frac{\partial^2 z}{\partial y \partial x} \qquad (18.4)$$

In other words the order of the differentiation does not matter. This implies that:

$$\left(\frac{\partial M}{\partial y}\right)_x = \left(\frac{\partial N}{\partial x}\right)_y \qquad (18.5)$$

This principle is now applied to an equation for the internal energy:

$$u = f(s,v) \qquad (18.6)$$

so that:

$$du = \left(\frac{\partial u}{\partial s}\right)_v ds + \left(\frac{\partial u}{\partial v}\right)_s dv \qquad (18.7)$$

But we also know that one version of the combined first and second laws is:

$$du = T\,ds - p\,dv \qquad (18.8)$$

Now:

$$\frac{\partial^2 u}{\partial s \partial v} = \frac{\partial^2 u}{\partial v \partial s} \qquad (18.9)$$

or:

$$\left(\frac{\partial T}{\partial v}\right)_s = -\left(\frac{\partial p}{\partial s}\right)_v \qquad (18.10)$$

This is the first Maxwell equation. A second one can be derived from another version of the combined first and second laws:

$$dh = T\,ds + v\,dp \qquad (18.11)$$

and is:

$$\left(\frac{\partial T}{\partial p}\right)_s = \left(\frac{\partial v}{\partial s}\right)_p \qquad (18.12)$$

Just as enthalpy was defined to be $u+pv$, so two more thermodynamic variables can be defined: the Helmholtz function a and the Gibbs function g:

$$a = u - Ts \qquad (18.13)$$

$$g = h - Ts \qquad (18.14)$$

These are the specific functions, that is, per unit mass. The units of each are J/kg. If required the symbols A and G would be used for the Helmholtz function and the Gibbs function for a mass m. The combined first and second laws can also be written in terms of a and g as follows:

$$da = -pdv - sdT \qquad (18.15)$$

$$dg = vdp - sdT \qquad (18.16)$$

From these equations two more Maxwell equations can be derived:

$$\left(\frac{\partial p}{\partial T}\right)_v = \left(\frac{\partial s}{\partial v}\right)_T \qquad (18.17)$$

$$\left(\frac{\partial v}{\partial T}\right)_p = -\left(\frac{\partial s}{\partial p}\right)_T \qquad (18.18)$$

Equations 18.17 and 18.18 are particularly useful because they contain entropy on only one side of the equation. They can therefore be used to eliminate entropy from thermodynamic equations.

18.3   Applications of the Maxwell equations

The Maxwell relations are very useful for a number of derivations and manipulations of thermodynamic equations. A number of examples are given here.

Example 18.1   Prove that the internal energy of an ideal gas is a function of temperature only.

Solution   This statement has been used a number of times previously, but has not yet been proved. The proof starts by writing the internal energy, a property of state, as a function of two other variables temperature and specific volume, hence:

$$u = f(T, v) \qquad (18.19)$$

and differentiating this:

$$du = \left(\frac{\partial u}{\partial T}\right)_v dT + \left(\frac{\partial u}{\partial v}\right)_T dv \qquad (18.20)$$

Now in eqn 18.20 it is known that:

$$\left(\frac{\partial u}{\partial T}\right)_v = c_v \qquad (18.21)$$

From one of the versions of the combined first and second laws:

$$du = Tds - pdv \tag{18.22}$$

$$\left(\frac{\partial u}{\partial v}\right)_T = T\left(\frac{\partial s}{\partial v}\right)_T - p \tag{18.23}$$

To prove that the internal energy is a function of temperature only, we need to show that the term on the left-hand side of eqn 18.23 is equal to zero. The awkward term involving entropy on the right-hand side of this equation can be removed by substituting from one of the Maxwell equations, eqn 18.17, to give:

$$\left(\frac{\partial u}{\partial v}\right)_T = T\left(\frac{\partial p}{\partial T}\right)_v - p \tag{18.24}$$

so that:

$$du = c_v dT + \left[T\left(\frac{\partial p}{\partial T}\right)_v - p\right] dv \tag{18.25}$$

From the ideal gas equation $pv = RT$:

$$\left(\frac{\partial p}{\partial T}\right)_v = \frac{R}{v} = \frac{p}{T} \tag{18.26}$$

so that:

$$T\left(\frac{\partial p}{\partial T}\right)_v - p = 0 \tag{18.27}$$

and then:

$$du = c_v dT \tag{18.28}$$

This means that for an ideal gas: $u = f(T)$ only.

Example 18.2   Prove that the enthalpy of an ideal gas is a function of temperature only.

Solution   The proof can be performed in an exactly similar manner to that for internal energy starting from the expression of enthalpy being a function of two state variables, temperature and pressure.

However, if it can be assumed that internal energy depends only on temperature, then it is much easier to write:

$$h = u + pv \tag{18.29}$$

and for an ideal gas:

$$h = u + RT \tag{18.30}$$

If u is a function of temperature only, so is $u + RT$. Therefore the enthalpy of an ideal gas is a function of temperature only.

**Example
18.3**

Derive a general expression for $c_p - c_v$ valid for any substance in terms of the temperature, specific volume, coefficient of thermal expansion and the compressibility.

Solution

The starting point for the derivation is to use $u = f(T, v)$ and $h = f(T, p)$ so that:

$$du = c_v dT + \left(\frac{\partial u}{\partial v}\right)_T dv \tag{18.31}$$

and:

$$dh = c_p dT + \left(\frac{\partial h}{\partial p}\right)_T dp \tag{18.32}$$

From the combined first and second laws:

$$du = Tds - pdv \tag{18.33}$$

$$dh = Tds + vdp \tag{18.34}$$

and combining eqns 18.33 and 18.34:

$$du + pdv = dh - vdp \tag{18.35}$$

Then substituting from eqns 18.31 and 18.32:

$$c_v dT + \left[p + \left(\frac{\partial u}{\partial v}\right)_T\right] dv = c_p dT - \left[v - \left(\frac{\partial h}{\partial p}\right)_T\right] dp \tag{18.36}$$

For the particular case of constant pressure ($dp = 0$), eqn 18.36 can be written:

$$c_v dT + \left[p + \left(\frac{\partial u}{\partial v}\right)_T\right] dv = c_p dT \tag{18.37}$$

and rearranging this equation gives:

$$c_p - c_v = \left[p + \left(\frac{\partial u}{\partial v}\right)_T\right] \left(\frac{\partial v}{\partial T}\right)_p \tag{18.38}$$

Now from $du = Tds - pdv$, and from the Maxwell equation 18.17:

$$\left(\frac{\partial u}{\partial v}\right)_T = T\left(\frac{\partial s}{\partial v}\right)_T - p = T\left(\frac{\partial p}{\partial T}\right)_v - p \tag{18.39}$$

and substituting from eqn 18.39 into eqn 18.38 gives:

$$c_p - c_v = T\left(\frac{\partial p}{\partial T}\right)_v \left(\frac{\partial v}{\partial T}\right)_p \tag{18.40}$$

This is a perfectly general equation for the difference in the specific heats for any substance. It can be made more useful by expressing it in terms of two properties commonly tabulated. These are the coefficient of thermal expansion β, and the compressibility κ:

$$\beta = \frac{1}{v}\left(\frac{\partial v}{\partial T}\right)_p \tag{18.41}$$

$$\kappa = -\frac{1}{v}\left(\frac{\partial v}{\partial p}\right)_T \tag{18.42}$$

After some manipulation, it can be shown that eqn 18.40 can be written in the form:

$$c_p - c_v = \frac{\beta^2 T v}{\kappa} \tag{18.43}$$

This is the required equation.

Example 18.4

Show that, for an ideal gas, eqn 18.43 produces the result:

$$c_p - c_v = R$$

Solution

From the ideal gas equation $pv = RT$, and from eqns 18.41 and 18.42:

$$\beta = \frac{1}{T} \quad \text{and} \quad \kappa = \frac{1}{p}$$

Substituting these results into eqn 18.43 gives:

$$c_p - c_v = \frac{pv}{T} = R$$

which is the expected result.

Example 18.5

Find the value of $c_p - c_v$ for copper at a temperature of 300 K for the following property values:

$$\beta = 51 \times 10^{-6} \text{ K}^{-1}$$
$$\kappa = 7.7 \times 10^{-9} \text{ m}^2/\text{N}$$
$$v = 1.12 \times 10^{-4} \text{ m}^3/\text{kg}$$

Solution

Substituting into eqn 18.43 gives:

$$c_p - c_v = \frac{(51 \times 10^{-6})^2 \times 300 \times 1.12 \times 10^{-4}}{7.7 \times 10^{-9}} = 0.01 \text{ J/kg K}$$

compared to the tabulated specific heat value of 380 J/kg K. The difference in the specific heat values for copper (and indeed for all solids and liquids) is minute compared with the specific heats themselves, and so only for gases are we concerned about different specific heats at constant pressure and at constant volume.

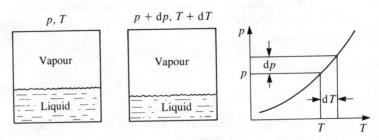

Fig. 18.1 Liquid and vapour in equilibrium in a closed box

18.4 The Clausius–Clapeyron equation

The Clausius–Clapeyron equation is concerned with the variation of saturation vapour pressure with temperature, and in practice is one of the most useful equations of thermodynamics. Figure 18.1 shows a closed box containing a single pure substance partly as vapour and partly as liquid. The two phases are in equilibrium. Consider now what happens if a very small amount of the liquid vaporizes. The system temperature and pressure are constant. From eqn 18.16 the change in the Gibbs function during this process must be zero as the pressure and the temperature do not change. This implies that the Gibbs function for saturated liquid and the Gibbs function for saturated vapour in equilibrium with the liquid are the same, that is $g_f = g_g$. This can be tested for water and steam at a pressure of 1 bar.

Example 18.6

Calculate the Gibbs function for saturated water and steam at 1 bar.

Solution

The values of the Gibbs function are not normally tabulated, and have to be worked out from $g = h - Ts$. The necessary values of the variables are:

$$h_f = 417.5 \text{ kJ/kg}$$
$$s_f = 1.3027 \text{ kJ/kg K}$$
$$h_g = 2675.4 \text{ kJ/kg}$$
$$s_g = 7.3598 \text{ kJ/kg K}$$
$$T_{sat} = 372.8 \text{ K (note: the } T_{sat} \text{ must be an absolute temperature)}$$

Then from $g = h - Ts$ for each phase:

$$g_f = 417.5 - 372.8 \times 1.3027 = -68.2 \text{ kJ/kg}$$

$$g_{\mathrm{g}} = 2675.4 - 372.8 \times 7.3598 = -68.2 \ \mathrm{kJ/kg}$$

The values of g_{f} and g_{g} are the same as expected. There is no particular significance to the negative sign.

If now the system pressure in Fig. 18.1 is increased slightly from p to $p + dp$, and the temperature increases from T to $T + dT$, then the Gibbs function of each phase changes slightly so that:

$$g_{\mathrm{f}} + dg_{\mathrm{f}} = g_{\mathrm{g}} + dg_{\mathrm{g}} \tag{18.44}$$

and therefore:

$$dg_{\mathrm{f}} = dg_{\mathrm{g}} \tag{18.45}$$

Now each of these dg terms can be expressed as $vdp - sdT$ so that:

$$v_{\mathrm{f}} dp - s_{\mathrm{f}} dT = v_{\mathrm{g}} dp - s_{\mathrm{g}} dT \tag{18.46}$$

and then rearranging eqn 18.46 gives:

$$\frac{dp}{dT} = \frac{s_{\mathrm{g}} - s_{\mathrm{f}}}{v_{\mathrm{g}} - v_{\mathrm{f}}} \tag{18.47}$$

Equation 18.47 contains the total differential dp/dT because, from the phase rule, the saturation vapour pressure p only depends on the saturation temperature T. The entropy difference in eqn 18.47 can be expressed in terms of an enthalpy difference:

$$s_{\mathrm{g}} - s_{\mathrm{f}} = \frac{h_{\mathrm{g}} - h_{\mathrm{f}}}{T} \tag{18.48}$$

so that eqn 18.47 can be written:

$$\frac{dp}{dT} = \frac{h_{\mathrm{g}} - h_{\mathrm{f}}}{T(v_{\mathrm{g}} - v_{\mathrm{f}})} \tag{18.49}$$

Equation 18.49 is the Clausius–Clapeyron equation. It can be further simplified by assuming that $v_{\mathrm{g}} \gg v_{\mathrm{f}}$. This is true for systems where the pressure is well below the critical pressure. If also the vapour behaves like an ideal gas, then $v_{\mathrm{g}} = RT/p$. With these assumptions and writing h_{fg} in place of $h_{\mathrm{g}} - h_{\mathrm{f}}$, eqn 18.49 becomes:

$$\frac{dp}{dT} = \frac{h_{\mathrm{fg}} p}{RT^2} \tag{18.50}$$

Alternatively, eqn 18.50 can be written as:

$$\frac{d(\ln p)}{d(1/T)} = -\frac{h_{\mathrm{fg}}}{R} \tag{18.51}$$

Equation 18.51 suggests that if h_{fg} (otherwise known as the latent heat of vaporization) is constant, then a graph of $\ln p$ against $1/T$

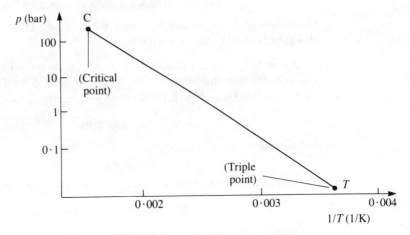

Fig. 18.2 Plot of $\ln p$ against $1/T$ for water

will be a straight line, as shown in Fig. 18.2 with a slope of $-h_{fg}/R$. Figure 18.2 is plotted for steam and water for a wide pressure range. At high pressure the line is not quite straight because:

1. h_{fg} is not constant: it falls as the pressure increases;

2. at high pressure the vapour does not behave ideally;

3. at high pressure it is no longer true that $v_g \gg v_f$.

However, the Clausius–Clapeyron equation, in the form of eqn 18.49 or of eqn 18.51, is a very useful way of interpolating to find the vapour pressure at intermediate pressure values in a table. It can also be used, with care, for extrapolation outside the range of tabulated data.

18.5   A simple derivation of Bernoulli's equation

Bernoulli's equation is one of the basic equations of fluid mechanics. It is usually derived by considering a small element of fluid, and then writing down the forces on that element. There will typically be a body force due to gravity, and a pressure force because the pressures acting on opposite faces of the element are not equal. The resultant force is then put equal to the mass of the fluid in the element multiplied by the acceleration of that fluid. This derivation is somewhat tedious, and a much simpler thermodynamic derivation is available.

This derivation starts with the steady-flow energy equation for a unit mass of the fluid:

$$Q - W_s = \Delta\left(h + \frac{c^2}{2} + gz\right) \tag{18.52}$$

Here c is the velocity of the fluid and z is the height above some arbitrary horizontal datum level. The differential form of eqn 18.52 is:

$$dQ - dW_s = dh + cdc + gdz \tag{18.53}$$

If it is now assumed that there is no shaft work, and that the change is reversible so that we can write $dQ = Tds$, then eqn 18.53 becomes:

$$Tds = dh + cdc + gdz \tag{18.54}$$

From the combined first and second laws, we know that a perfectly general equation is:

$$dh = Tds + vdp \tag{18.55}$$

Now combining eqns 18.54 and 18.55:

$$vdp + cdc + gdz = 0 \tag{18.56}$$

If the specific volume is constant, and putting $v = 1/\rho$, integrating eqn 18.56 leads to:

$$\frac{p}{\rho} + \frac{c^2}{2} + gz = C \tag{18.57}$$

where C is a constant. Equation 18.57 is Bernoulli's equation. The conditions for it to apply are as follows.

1. The flow must be steady. If it was not steady we could not have started with the steady flow energy equation.

2. The flow must be reversible since we have used $dQ = Tds$. This implies that there must be no friction between the fluid and any containing walls. Real flows are never exactly reversible.

3. There must be no shaft work since we have put $dW_s = 0$.

4. The fluid must be incompressible, in other words the density ρ must be constant. This fact was used in the final integration to give eqn 18.57.

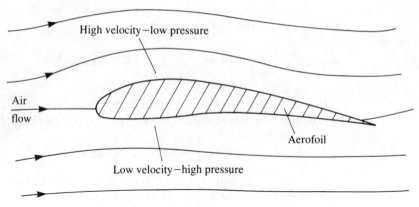

Fig. 18.3   The flow around an aerofoil

Note that it is not necessary to specify that the heat transfer is zero, only that it be accomplished reversibly.

From Bernoulli's equation the pressure distribution around an aerofoil[1] can be deduced. Figure 18.3 shows the flow around a typical aerofoil. The air flowing over the top surface has to travel further than that flowing over the bottom surface. Therefore the velocity over the top is higher, and so from Bernoulli's equation the pressure at the top of the aerofoil must be lower than the pressure at the bottom surface of the aerofoil. This difference in pressure causes a net resultant force upwards. This is the lift produced by the aerofoil.

Example 18.7

A Boeing-747 airliner has a wing area of 500 m², and cruises at 950 kilometres per hour (264 m/s) in level, steady flight. The air pressure is 0.36 bar, and the temperature is 236 K. If the air velocity over the lower surface of the aerofoil is equal to the cruising speed, and the velocity over the upper surface is 15% greater, find the mass of the airliner.

Solution

First calculate the air density from the ideal gas equation:

$$\rho = \frac{p}{RT} = \frac{0.36 \times 10^5}{287 \times 236} = 0.53 \text{ m}^3/\text{kg}$$

From Bernoulli's equation the pressure difference between the lower and upper surfaces of the aerofoil is:

[1] An aerofoil is an aircraft wing cross section.

$$\Delta p = p_l - p_u = \frac{\rho}{2}(c_u^2 - c_l^2) = \frac{0.53}{2}(264^2 \times 1.15^2 - 264^2) = 5956 \text{ N/m}^2$$

Here the subscript l denotes the lower surface, and the subscript u the upper surface. The lift is this pressure difference multiplied by the wing area. In steady, level flight the lift force is equal to the weight of the airliner. So the mass m of the airliner is:

$$m = \frac{\Delta p \times \text{wing area}}{g} = \frac{5956 \times 500}{9.81} = 304000 \text{ kg} = 304 \text{ tonne}$$

18.6 Problems

18.1 Prove that for an ideal gas:

$$g_T = g_T^0 + RT \ln \frac{p}{p^0}$$

where:

g_T = Gibbs function at T and p
g_T^0 = Gibbs function at T and p^0

Taking $T = 373$ K and $p^0 = 1$ bar, calculate g_T^0 for steam using data from steam tables. Then using the above equation calculate g_T for steam at $p = 0.1$ bar, and for $p = 0.01$ bar.

18.2 Prove that for a liquid like water the following equation is approximately true:

$$g_T = g_T^0 + v(p - p^0)$$

where:

g_T = Gibbs function at T and p
g_T^0 = Gibbs function at T and p^0
v = specific volume of the liquid

Hence taking $T = 373$ K calculate g_T for water at $p = 10$ bar, and for $p = 100$ bar.

18.3 From the results of questions 18.1 and 18.2 sketch a graph of the variation of the Gibbs function with pressure for water substance (water or steam) at 373 K.

18.4 Benzene (C_6H_6) boils at 1 bar at 80.1°C and has a latent heat of vaporization of 403 kJ/kg. Estimate the boiling point of benzene at 5 bar.

18.5 The following table gives values of the boiling temperature of toluene at various pressures. One of the temperature entries in this table is wrong. Correct this entry, and estimate the latent heat of vaporization per kg-mole.

Boiling point (°C)	Pressure (bar)
110.6	1
136.5	2
178.0	5
227.1	10
262.5	20
319.0	40

18.6 Prove the following general thermodynamic equations:

(a)
$$\left(\frac{\partial u}{\partial p}\right)_T = -T\left(\frac{\partial v}{\partial T}\right)_p - p\left(\frac{\partial v}{\partial p}\right)_T$$

(b)
$$c_p = T\left(\frac{\partial v}{\partial T}\right)_p \left(\frac{\partial p}{\partial T}\right)_s$$

(c)
$$ds = \frac{c_p}{T}dT - \left(\frac{\partial v}{\partial T}\right)_p dp$$

(d)
$$\left(\frac{\partial T}{\partial p}\right)_h = \frac{1}{c_p}\left[T\left(\frac{\partial v}{\partial T}\right)_p - v\right]$$

19

Solutions to numerical problems

1.1 -40°F = -40°C

1.2 2141

1.3 1.9 K

1.4 –

1.5 202 s

1.6 80 s

2.1 22.5% Cl^{35}

2.2 35.7 kg; oxygen 1.48×10^{26}; nitrogen 5.93×10^{26}

2.3 24.9 m^3; 12.4 kg; 931 K

2.4 Calculated values (kg/m^3): 0.068; 37.1; 0.28; 28.0;
     Values from steam tables (kg/m^3): 0.068; 55.4; 0.28; 30.5.

2.5 353.5 K (80.5°C)

2.6 0.89 bar; 0.32 bar; 1.1 Pa

2.7 1.7×10^{44}

3.1 m^5/kg s^2; m^3/kg

3.2 0.375

3.3 300 K; 73 bar; 2.91×10^{-3} m^3/kg

3.4 –

3.5 At values of v_r of 3, 1, and 0.5 the values of p_r are 0.667, 1, and 4 respectively, and the corresponding values of Z are 0.75, 0.375, and 0.75.

4.1 0.235

4.2 –

4.3 99.63°C, 0.590, saturated; 400°C, superheated; 2 bar, superheated; V and x indeterminate, saturated; 118 kg, 99.63°C, saturated

5.1 60, 0, 40, 100%; 60, 80, 120, 33%

5.2 36%; 20%

6.1 3.5 bar; 365.2 K

6.2 48.8 J; 246.4 J, 295.2 J

6.3 6250 kJ; -3125 kJ; 3125 kJ

6.4 5925 kJ; -8527 kJ; -2602 kJ

6.5 –

6.6 111 MJ

6.7 507 MJ; 401°C

6.8 50.65 bar; 365.6°C

6.9 2932 J; 471 J

7.1 268.9 K

7.2 131 kJ/kg

7.3 99.63°C, $x = 0.326$; 140°C, superheated; 500°C

7.4 –

7.5 0.267; 0.9986

7.6 0.138

7.7 10.67 kg; 443 K

7.8 6.08 kg; 237°C

9.1 250 J, reversible; 250 J, impossible; <250 K, 300 J; -1800 J, irreversible; 2000 J, 3000 J

9.2 –

9.3 0.005 kJ; 206.9 kJ

9.4 931.6°T; 12.0 K

10.1 -193 J/K; -1014 J/K (tables -1011 J/K); -243 J/K; -173 J/K; -8383 J/K

10.2 20.0 J/kg K; 0.2732 kJ/kg K; 6.4 J/kg K; 189 J/kg K; 0.282 kJ/kg K

10.3 $1.3475T$; $3.305T$

10.4 1010 kW, 500.5 K; 771 kW

10.5 328 K; 380 m/s, zero

10.6 400 K; 199 J/kg K

10.7 X → Y

10.8 55.4 kJ/kg; 136.2 kJ/kg; 239.8 kJ/kg

11.1 –

11.2 411 J/kg K; 294 J/kg K

11.3 14.5 kJ/kg K; 10.4 kJ/kg K

11.4 –

11.5 16.6 kJ/kg K

11.6 1.00; 1.22

12.1 0.307

12.2 –; $v_3/v_2 = 5.66$; $4.66v_1$; $0.16p_1$

12.3 –

12.4 0.088

12.5 $0.31p_1$

12.6 $\eta_{th} = 1 - \frac{T_l}{T_h}$; $0.632p$; $2.56p$; $0.894p$

12.7 –

12.8 3.3 bar

12.9 37 bar; 1100 kJ/kg

13.1 164°C; 338 kJ/kg

13.2 ideal gas behaviour: 213°C; 574 kJ/kg

13.3 steam table results: 220°C; 565 kJ/kg

13.4 ideal gas behaviour: -1°C; 1002 kJ/kg; steam table results: 30°C (dryness = 0.96); 1024kJ/kg

13.5 dryness = 0.77; 1395 kJ/kg

13.6 dryness = 0.90; 1400 kJ/kg

13.7 89%; single stage 71%

13.8 109°C; 269 kJ/kg

14.1 37.3%, $x = 0.69$, 1.00×10^{-3} kg/kJ

14.2 –

14.3 33.5%, $x = 0.73$, 1.12×10^{-3} kg/kJ

14.4 36.1%, $x = 0.87$, 8.36×10^{-4} kg/kJ

14.5 38.1%, $x = 1.00$, 7.00×10^{-4} kg/kJ

14.6 40.1%, $x = 1.00$, 7.60×10^{-4} kg/kJ, 0.188

14.7 –

15.1 56.5% for $\gamma = 1.67$; 38.1% for $\gamma = 1.30$

15.2 501 K; 599 K; 201 kJ/kg; 501 kJ/kg; 40%

15.3 536 K; 639 K; 126 kJ/kg; 466 kJ/kg; 27%

15.4 yes; 50%; 35%

15.5 –; 279 kJ/kg; 3108 kJ/kg; 11.12; –

15.6 25°C

16.1 1.5 bar; 4.57; 29 kJ/kg; 56 mm

16.2 4.08; 2213 kJ/kg; 45 mm

16.3 4.17; 1121 kJ/kg; 8.4 mm

16.4 –

17.1 538 K, 3338 K, 1863 K; 12.6 bar, 77.9 bar, 6.2 bar; 2800 kJ/kg, 1563kJ/kg; 44.2%; 16.0 bar

17.2 –

17.3 44.2%; 33.6 bar

17.4 –

17.5 1.875×10^{-8} m^3; 38.7:1

18.1 -69.7 kJ/kg; -466 kJ/kg; -863 kJ/kg

18.2 -68.8 kJ/kg; -59.4 kJ/kg

18.3 –

18.4 142.5°C

18.5 227.1°C should be 216°C; 34 MJ/kg-mole

18.6 –

Index